W9-BAQ-828

ADVENTURES IN
OCEAN EXPLORATION

ADVENTURES IN OCEAN EXPLORATION

From the Discovery of the Titanic to the Search for Noah's Flood

Robert D. Ballard
with Malcolm McConnell

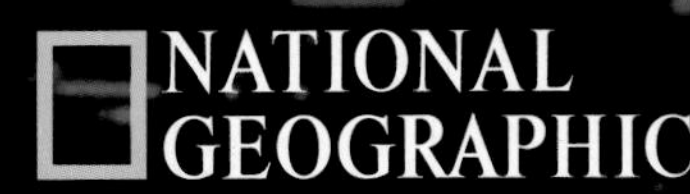

Washington, D.C.

CONTENTS

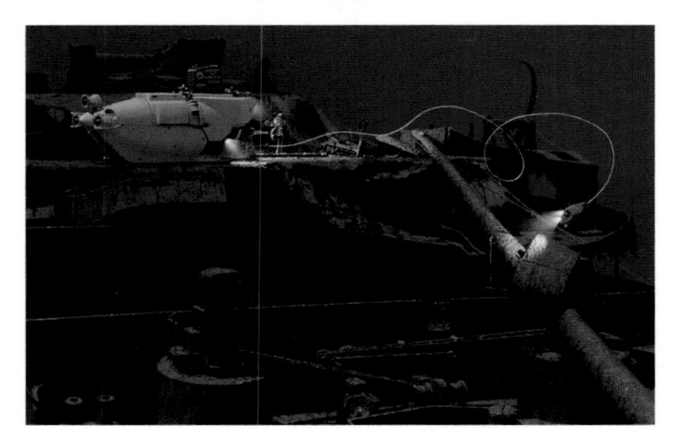

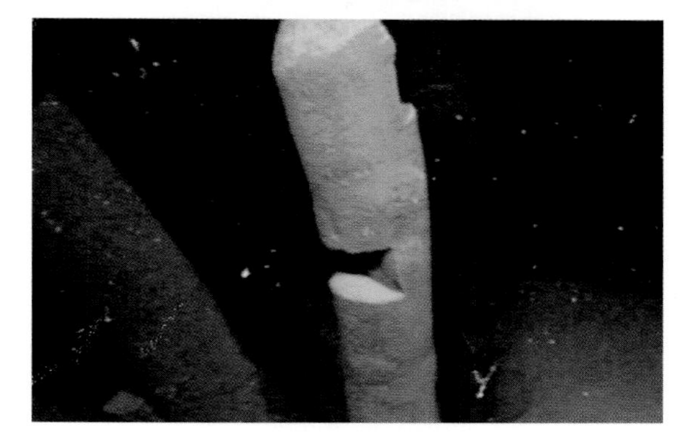

Preceding pages: Page 1: Alvin *sinks at 30 meters a minute on the Cayman Expedition, carrying two scientists and a pilot within its titanium-alloy sphere.*

Pages 2-3: A submersible allows two passengers to inspect the Cayman Wall. The pilot navigates this craft to depths up to a thousand feet for up to eight hours.

Introduction

In this allegorical painting, the mythical Bronze Age Jason and the Argonauts land at Colchis on the present-day Black Sea in their quest for the Golden Fleece. To retrieve the treasure, Jason and his brave crew—like modern explorers—had to overcome monsters and natural obstacles on their hazardous journey.

I AM AN OCEAN EXPLORER AND A SCIENTIST. Since my first cruise as a high school student in 1959, I've been on 110 expeditions. I have explored the ocean's great depths in submersibles, visited lost liners on the sea bottom, and surveyed ancient shipwrecks.

I often meet people who comment on how "exciting" my work must be. That's certainly true. The joy of discovery is uniquely exciting. As a geologist on deep-ocean expeditions in the 1970s, I helped confirm the revolutionary theory of plate tectonics that changed the way we view our dynamic planet. On the expeditions that discovered the seafloor hydrothermal vents of the Pacific, our team found bizarre new creatures, which transformed the photosynthesis-based paradigm of the chain of life. But modern, scientific exploration is also beset with hard work and worry about lost, broken, or mysteriously malfunctioning equipment, schedule problems, possibly flawed hypotheses, bad weather, and a host of other difficulties that can cripple an expedition. Rarely these days do I find myself in a truly dangerous situation under the sea, mainly because I now do most of my exploration from the surface with sonar and ROVs, but I've certainly had my brushes with death in the crushing pressure of the black abyss.

Today, I am an Explorer-in-Residence at the National Geographic Society and president of the Institute for Exploration in Mystic, Connecticut. I spend part of each year on expeditions and part on the JASON Project, an interactive, multimedia international "classroom" that links hundreds of thousands of students in grades four to nine with scientists working at remote sites. It is in the eager faces of the children who actually work with my colleagues at the project sites that I see a reflection of myself as a boy.

I grew up in the San Diego suburb of Pacific Beach, so it's no surprise that I was attracted to the sea at a young age. But while other boys were absorbed by the Navy's big gray aircraft carriers and cruisers steaming in and out of the harbor, I could be found exploring the tidal pools of Mission Bay, California. Everything on top of the water seemed boring. I always wanted to stick my head beneath the surface. Until I got a diving mask, I had to make do with glass bowls from my mother's kitchen to catch blurry glimpses of the creeping hermit crabs and the rippling yellow crowns of the sea anemones. There was an entire secret world under the sea, and I knew before I was ten that somehow I would be part of it.

I became engrossed with submarines, machines that could carry people into that secret world. The model subs I built of mason jars and tin cans worked well enough in the tidal pools, but I wasn't satisfied with toys. I read and reread Jules Verne's *Twenty Thousand Leagues Under the Sea* and became obsessed with his futuristic *Nautilus*, and with exploring the great ocean depths.

But my childhood dreams of deep-sea exploration took decades to reach fruition. I earned my Ph.D. in geology and then served a demanding apprenticeship at the Woods Hole Oceanographic Institution (WHOI) on Cape Cod. For me, the 1970s were a period of rigorous scientific exploration, focused on plate tectonics and hydrothermal vents. In the 1980s, I turned to undersea archaeology and maritime history, endeavors which proved quite successful, most notably on the 1985 joint U.S.-French expedition that discovered the *Titanic*. Over the years I've learned that the deep sea is a vast museum, its frigid darkness preserving many of the priceless artifacts that reach the bottom through mishap or act of war.

In the 50 years since I probed the shallows of Mission Bay, I have often pondered the distinctly human impulse to explore. And I have studied the lives of history's most notable ocean explorers, discovering that we share common attributes. Our lives have all involved what I have come to call the "Cycle of Exploration"—the relentless urge to journey forth seeking knowledge, then to return home with that precious cargo to light the spark of exploration in others.

In Western civilization, this cycle is rooted in ancient myths and legends such as Jason and the Argonauts, and what was immediately relevant to the Minoans voyaging north into the forbidding frontier of the Black Sea remains pertinent to 21st century explorers. Indeed, the cycle of exploration has endured almost intact over the last three millennia.

These elements comprise the cycle:

The Vision: The primal impulse to explore—ancient people's driving curiosity to discover, often spurred by myths of fabulous wealth, the Golden Fleece that Jason and his brave sailors sought. Our contemporary analog of the vision or dream is the scientific hypothesis.

Preparation: Gathering the crew of voyagers or a scientific expedition.

Journeying Forth: Setting out into the unknown. All explorers from the mythical Jason, to Columbus, to my oceanographer colleagues reach a point where they must actually commit themselves.

Overcoming Obstacles: For centuries, seafaring explorers faced "monsters" (unknown creatures such as whales), mutiny by dispirited crew, and devastating hardships and illness, including starvation and scurvy. Quick wits and improvisation has long been the explorer's friend. Today, scientific explorers must often overcome the obstacle of staid, inaccurate conventional wisdom.

Discovering Truth: The Golden Fleece, a new continent, or, in the 1970s, the true nature of the Mid-Atlantic Ridge (which confirmed the theory of plate tectonics). **Sharing the Truth:** The return of the heroic voyager (Jason, Ulysses, and other mythical explorers). In our day, scientific explorers share their findings in peer-reviewed journals, symposia, and publications such as NATIONAL GEOGRAPHIC. **Sparking New Vision:** Minoan and Homeric heroic voyage myths spurred Greek colonization of the Black Sea and Mediterranean shores. Contemporary scientific exploration almost always engenders more questions than are answered, thus opening windows for additional hypotheses. Each year, the JASON Project sparks the quest for knowledge in hundreds of thousands of schoolchildren.

Beyond the unquenchable curiosity inherent to the cycle of exploration, I've come to realize that my modern colleagues share with our predecessors other essential qualities: determination, patience, optimism, physical endurance, and above all, perseverance.

All modern explorers owe a great debt to those who have come before us. While cruising the South Pacific, for example, I have entered coral-fringed anchorages that were first charted by British Royal Navy Captain James Cook in the 1700s. As a 21st century explorer, I cannot define where the questing voyages of discovery across the Atlantic that the Vikings undertook a thousand years ago or those epic journeys of the Phoenician seafarers 1,800 years before the Norsemen end and my profession begins. It is an unbroken line. So I always feel a stab of anguish when I first see a shipwreck—no matter how far removed in time—appear from the drifting haze of sediment in the sensitive video eye of our faithful ROV *Jason*.

So I decided to write this history of ocean exploration. I did not attempt an exhaustive academic study. Instead, I tried to weave together enough information about seamanship, navigation, shipbuilding, and propulsion—as well as the dark side of ocean exploration, war at sea—to give the reader a good understanding of how these skills evolved over the centuries to permit human beings, who were landlocked for most of their existence, to become masters of a planet whose surface is 71 percent water. It is that earlier seafaring technology and expertise that have allowed ocean explorers of my generation to develop much more advanced tools, such as precision sonars and remotely operated vehicles (ROVs), with which we can visit deep-sea wrecks and archaeological sites from a shipboard laboratory.

I think the best way to tell this story is the way that I learned it, through the expeditions that have taken me to the ocean's depths and the world's remote islands, often following the ghostly trail of great past explorers.

I The Ancient Depths

Preceding pages: **An eel undulates across the stacked amphorae in the hold of *Elissa*, one of two Iron Age Phoenician round boats my expedition discovered in June 1999, the oldest shipwrecks yet found in the deep sea.**

The archaeological dig at Ashkelon, the ancient Philistine port north of Gaza in present-day Israel, presents a fascinating window on the past. For millennia, Ashkelon was an important transshipment center and port of call for far-ranging Phoenician sailors.

Jason's powerful floodlights cast a chalky glare across the sandy bottom of the Mediterranean 1,300 feet (400 meters) below. It was after midnight on Friday, June 11, 1999. Our expedition's chartered ship, the British offshore supply vessel *Northern Horizon*, used its dynamic thrusters linked to global positioning system satellites to hold a precise surface position above the seafloor, 60 miles off the Israeli city of Ashkelon, an ancient Philistine port north of Gaza. Linked to the ship by a fiber-optic tether, *Jason* crept cautiously forward above the bottom, guided by the expert hands of pilot Will Sellers who delicately worked the ROV's joystick in the shipboard control van.

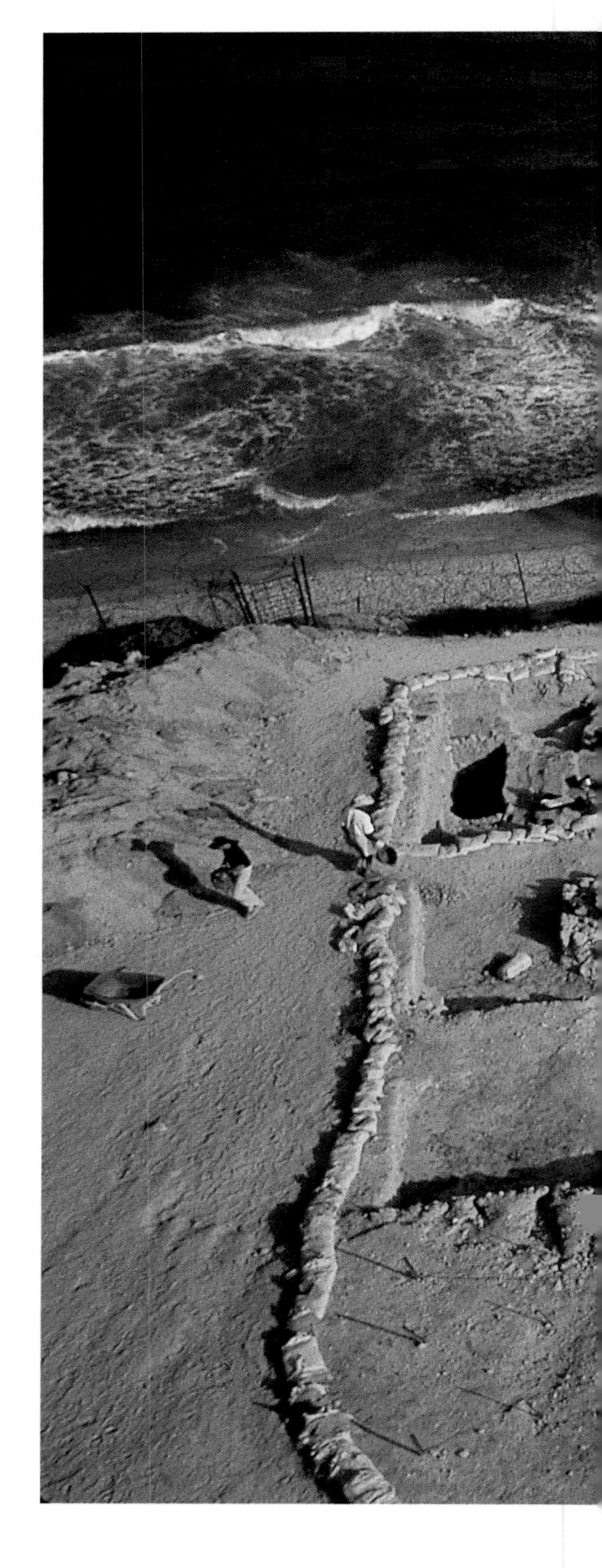

I stood in the blue glow of the video monitors and computer screens, waiting tensely as the digits winked on the wide navigation status display screen that indicated the closing sonar range to the target. We were hunting ancient shipwrecks, possibly the oldest ever discovered in the deep sea. Two years before, during a summer 1997 Mediterranean cruise, the U.S. Navy's small nuclear submarine *NR-1* had unsuccessfully searched these waters for the sunken Israeli submarine, *Dakar*, which had disappeared with all hands in 1968. Although the *NR-1* had not found that lost sub, the *NR-1*'s powerful sonar had identified several intriguing targets and her video cameras had taped grainy images of what appeared to be cylindrical objects on the bottom.

Because I had dived aboard the sub earlier that summer in a very successful exploration of Roman wrecks near Sicily, the crew showed me their mysterious tapes from the eastern Mediterranean off Ashkelon. Could these be more wrecks dating back to Romans times or even earlier? Archaeologist George Bass had discovered the wreck of a small, coast-hugging Minoan trading vessel from 1300 B.C. near the Turkish port of Fethiye in the 1970s. But that was in relatively shallow water.

The *NR-1*'s possible wrecks were in deep water, well offshore. If they actually were ancient ships, they might mark long-forgotten Phoenician trade routes. The mysterious Phoenicians, who settled the Levant (modern Lebanon) as early as 3000 B.C., were the most adventurous explorers and seafarers of the ancient Middle East. Although their language is not well understood, they seem to have called themselves the *Kena'ani*—in Hebrew, Canaanites. Their principal occupation seems to have been maritime commerce. Phoenician trading settlements have been identified along

Harvard University archaeologist Dr. Lawrence E. Stager (center) leads the 1999 dig at Ashkelon. He departed land that summer to join the underwater search for the Phoenician ships.

the coast of the eastern Mediterranean. The Phoenicians also established colonies—Carthage near present-day Tunis being the most famous—in Cyprus, North Africa, and far-off Spain.

Although the Phoenicians were the most notable seaborne traders of the late Bronze Age and early Iron Age, archaeologists knew relatively little about their actual ships and Mediterranean trade routes. How did they stow their cargoes for maximum stability? Did they resort to long sweeping oars when the wind failed? And, if so, would a hull design that accommodated such oars have survived the millennia on the seafloor? Did they always sail within sight of the land, skipping from one settlement to another, or did they beach their small ships at night?

I am an oceanographer and a geologist by profession, not an archaeologist. So

I turned to Harvard University's world-renowned expert in the ancient Middle East Dr. Lawrence E. Stager for answers. Larry, a bearded, jolly, bear of a scholar with scores of land expeditions behind him, possessed encyclopedic knowledge of the ancient artifacts he had extracted with great care from the dusty tells and wadies of the region. He could glance at a tiny sherd of painted pottery and tell you if it was Philistine, Assyrian, or Babylonian.

But when we viewed the *NR-1* tapes, Larry Stager was in a quandary. He was virtually certain that these shadowy objects were terra-cotta amphorae, but he was much less positive about their date. "It would be wonderful if they're really old, Bob," he had told me, studying a ghostly blue freeze-frame image. "But there's no way to know until we see them up close." Larry explained that the system of dating an amphora depended on its general shape, combined with such distinctive details as its handles, firing technique, neck proportions, and the lips of its opening. In the blurry videotapes, the cylinders could have been Iron Age Phoenician amphorae from the eighth or seventh century B.C., fifth century A.D. amphorae from the late Roman Empire, or even Byzantine jars from seven centuries later.

Naturally, we hoped the *NR-1*'s tapes revealed Phoenician artifacts because, if they did, we stood a reasonable chance of discovering the oldest shipwrecks yet found in the deep sea. And, if the wrecks were Iron Age, finding them would be true discovery, since there was no surviving record of such ships having gone missing. My expeditions that found the wreckages of the *Titanic* and the German battleship *Bismarck*, as difficult as they had proved to be, had actually been relocations, since the sinkings had been historic events that occurred at well-charted coordinates. Employing sonar to survey and discover the actual wreck sites, then using an ROV for archaeological exploration of literally uncharted Phoenician shipwrecks preserved in the deep sea would be a first for all of us. That exciting prospect had helped secure support for our expedition from the National Geographic Society, the Office of Naval Research, and Leon Levy Expeditions.

We had arrived on the site 36 hours earlier and conducted a sonar search of the three targets that *NR-1* had tentatively identified two years earlier. Even though an electrical problem with the ship's winch had temporarily looked like a "show stopper," as my *Titanic* expedition colleague, WHOI's Deep Submergence Laboratory's Dana Yoerger, put it, he and the crew managed to shift to an auxiliary generator.

In any event, the winch glitch on *Northern Horizon* was nothing like the incident that almost shut down our search for the *Titanic* in 1985. Then, we'd been towing the camera sled *Argo* near the sea bottom, 12,000 feet below, when the cable became fouled over the winch drum. The tension had increased to 17,000 pounds, rising toward the 20,000-pound breaking point as we scrambled on the stern of the expedition ship *Knorr* to rig a scaffold to ease the strain on the tow cable before it snapped and we lost our only search tool. Had the cable parted under that strain while we were working, any one of us could have been cut to pieces. But we managed to save our expedition during that frightening night on the Atlantic.

Since the 1985 *Titanic* uncovering, I had been on scores of expeditions with Dana and his team. Their DSL 120 side-scan sonar was state-of-the-art, using high frequency acoustic signals to produce detailed colored three-dimensional images. But in the initial, wider search mode, the sonar's blue-gray display screen simply revealed three

Will Sellers helps prepare the towed camera sled *Medea* during the 1999 expedition off Ashkelon. Our remotely operated vehicle (ROV) *Jason* always works tethered to *Medea*; *Medea* offers a birds-eye view of the larger vehicle's location at the exploration site.

smudgy wedges that looked like spilled coal dust. Those were our targets.

Now *Jason* was closing in on the first of these targets, which we had logged with the prosaic expedition name of "AA." As always when the point of discovery approached, the expedition control van became crowded. I stood beside Larry Stager watching the video screens and display panels as Will Sellers deftly tweaked the joystick's multiple buttons—which reminded me of a fighter pilot's complex controls—to ease the ROV down to a depth of 1280 feet (390 meters) on a northwest heading of 301 degrees, 203 feet (62 meters) above the bottom. Now he used the robot arm to drop two small ballast weights and the ROV descended on an even trim, thrusters driving it forward toward the target.

With a skilled pilot at the joystick, *Jason*'s seven computer-controlled thruster motors could maneuver the ton-and-a-half displacement ROV in millimeter increments. In neutral buoyancy trim, *Jason* could hover above the most vulnerable archaeological site, its video and still cameras, as well as its electronic imaging system, capturing a precious record before any artifacts were disturbed. When it came time to recovering artifacts, *Jason*'s pilots could use the robot arm with its articulated hydraulic hand to safely retrieve extremely fragile objects. I was convinced we had the ideal tools for this expedition in the combination of the DSL 120 side-scan sonar and the ROV *Jason*.

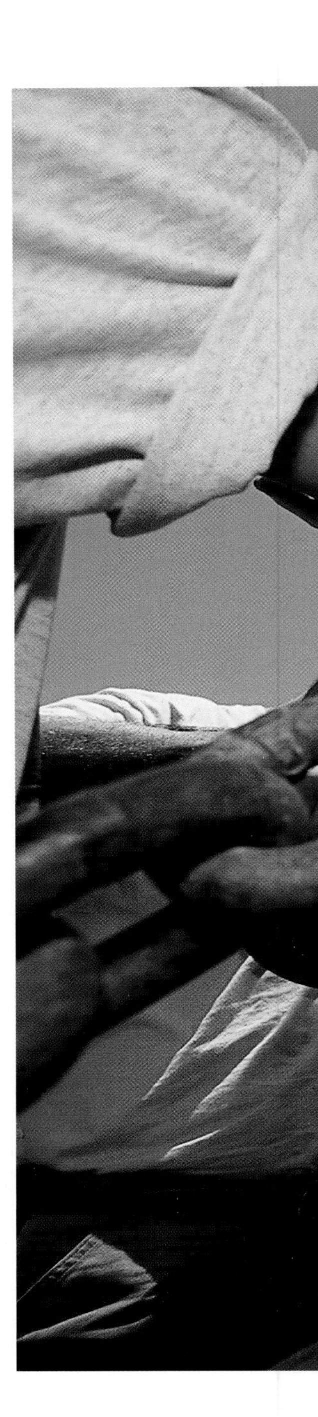

Another screen showed the wavering pattern of *Jason*'s close-range on-board sonar picking up a distinctive target that matched the DSL 120 image. I squeezed Larry Stager's shoulder. In the electronic glow of the monitors, his face was tight with anticipation. *Jason* was just a few meters above the bottom, moving silently forward, its floodlights cutting the darkness, startling small fish and pale, skittering crabs on the sand bottom.

Then I saw angular shapes. "That's not geology," I said, a bit of unconscious hubris I'd used on past expeditions to break the tension, the wise old geologist assuring the gang we were looking at artifacts, not natural features of the seafloor. The video image became sharper. Something glinted in the floodlight. "There it is," I said, somewhat less certainly now, "whatever it is, straight ahead."

The video monitor filled with an old-fashioned steel anchor and a rusty chain trailing off into the gloom. There were soft groans among Larry's Harvard team and a sharp sigh from Dr. Shelly Wachsmann of the Institute of Nautical Archaeology at Texas A&M University who stood nearby. He was an international expert on ancient ships and hoped to observe well-preserved Iron Age wrecks on this expedition. But all of us realized that the steel anchor and its chain came from an 18th or 19th century sailing vessel, not a Phoenician "round boat" from the eighth century B.C.

As expedition leader, I gave the order to raise *Jason* off the bottom and proceed

north toward the second target, which we had logged in as "AC." Although the distance between the two targets was only about one-and-a-half nautical miles, it would take around three hours to get the ROV redeployed for the close-in search. And those were anxious hours. The image of that steel anchor and chain bothered me. What if the other targets also turned out to be modern wrecks, Byzantine or even later? I'd found funding to bring 49 scientists and graduate students on this cruise. There were 55 tons of scientific equipment aboard *Northern Horizon*. What if we came up empty-handed?

I waited with Larry Stager trying to keep the conversation away from worst-case scenarios. But we were both worried. Having cruised the eastern Mediterranean for years, I knew the bottom was littered with modern wrecks, many the ubiquitous Greek and Cypriot sailing *caiques* that plied their trade between North Africa, Asia Minor, and Greece throughout most of the 19th and 20th centuries. Until the advent of plastic, many of these wooden merchant vessels carried their cargoes in large ceramic jars the modern Greeks call *qupia*. Was it possible the *NR-1* images were of such mundane transport jugs? If so, we'd find out when *Jason* inspected the next target.

The control van was even more crowded as *Jason* closed on target AC. There were people wedged in here from off-duty watches who should have been getting their sleep for the busy shifts ahead. But I didn't have the heart to discourage them. Once more, the ROV's scanning sonar acquired a nicely reflective image at a range of 492 feet (150 meters). Will Sellers drove *Jason* smoothly forward, high enough that the thrusters did not kick up sediment clouds and ruin visibility. On the sonar screen, the image appeared as an oval mound within a depression 49 feet (15 meters) long.

Flecks of sediment drifted across the floodlit video monitor as the sandy bottom unrolled toward us. Once more I felt Larry's shoulders tense. The screen suddenly filled with a great, heaped mound of tightly stacked amphorae. There were hundreds of them marching down the length of the wreck site in the long oval.

Whoops of triumph filled the van.

"Thar she blows," I said. From what I could see, there was nothing modern about these terra-cotta jars. On previous underwater archaeology expeditions, I had learned that the exposed hull planks of ancient wrecks were eventually devoured by the sea's voracious worms and wood-boring organisms. Only wooden hulls buried deeply enough to be protected from current scour and exposure to oxygenated seawater might survive. So, looking at a wreck such as this, in which the inorganic terracotta amphorae cargo had remained intact but the hull exposed above the mud line had vanished, it certainly seemed that we had discovered an ancient ship, not a 20th century Cypriot caique blown off course and capsized by a fierce sirocco.

I shifted my gaze from the video monitor to study Larry's expression. He was transfixed and spoke in awe as *Jason* passed slowly over the wreck. "That's the mother of all ships," he murmured.

Now I looked back at the shimmering image, suddenly no longer in the van, but virtually hovering a few meters above the wreck, my eyes absorbing what the video camera saw. This was the precious sense of "telepresence" that I had struggled with my colleagues from Woods Hole's Deep Submergence Laboratory to achieve over the previous two decades. I had the distinct, but eerie impression that *Jason* was showing us a ship made of glass, the ranks of fired-clay jars defining the vessel's shape and size.

As Will worked the thruster controls, the full shape and dimension of the ship became obvious, an elongated oval with dark, densely stacked amphorae marking the cargo holds and a looser collection of artifacts at bow and stern. Larry's face was glowing with a broad smile. "There's nothing bigger," he said, beaming at the monitor. "This is the first Iron Age ship to be found in the deep sea."

His colleague, Ph.D. candidate Susan Cohen, used a ruby red laser pointer to highlight a doughnut-shape on the monitor. "We've got the anchor," she proclaimed.

"Yes," Larry confirmed, "a stone anchor." He was almost overcome by profound emotion. "This is a night to remember. When you have this kind of moment, you never forget it."

Larry did not exaggerate. His initial assessment of this ship as Iron Age Phoenician placed it in the time of Homer, the Greek bard who created the epic tales of Western civilization, *The Iliad* and *The Odyssey*. Human eyes had not seen these stacked amphorae in almost 3,000 years.

Jason hovered near the stern and shifted its video eye toward two small, squat

As the Phoenician shipwreck *Tanit* grows larger in *Jason*'s video monitor, our expedition crew, Andy Bowen (right), Dr. Larry Stager (center), ROV pilot Martin Bowen (with joystick), and I share the wonder of discovery.

terra-cotta objects that Larry quickly identified as cooking pots. They marked the crew quarters and galley. Staring at the image of the bulbous pots, Larry became pensive. "I do think about the people who went down," he admitted.

So did I. Even though *Jason* had been on the wreck for less than an hour, I'd already reached tentative conclusions about the tragedy that had sunk this ship and taken the lives of those Phoenician sailors almost 3,000 years ago. First, the ship had not capsized, rolling over to dump its cargo of amphorae and heavy stone anchor before sinking. The vessel had gone down on an even keel and come to rest on the bottom horizontally, almost as if sailing underwater. That probably meant the vessel had been swamped by a huge storm wave while running before the wind, perhaps during one of the fierce spring or autumn gales that still plague this corner of the Mediterranean.

With the hull suddenly flooded, the Phoenician ship had settled evenly beneath the surface, reached the maximum descent velocity, given its horizontal position, then buried deep in the soft bottom sediments. Over the centuries, scouring currents had swept away the mud, exposing the upper hull, which fell prey to the wood-eaters. Only the stone anchor and heaped terra-cotta cargo and cooking pots remained in the oval depression of the wreck site.

But was this ancient ship an anachronism? We were miles from land. How often did Phoenician sailors venture from the comforting sight of the desert coast as they plied trade routes between their principal entrepôt port of Tyre and their colonies such as Carthage? Did they try to beach their ships at night, as some scholars have

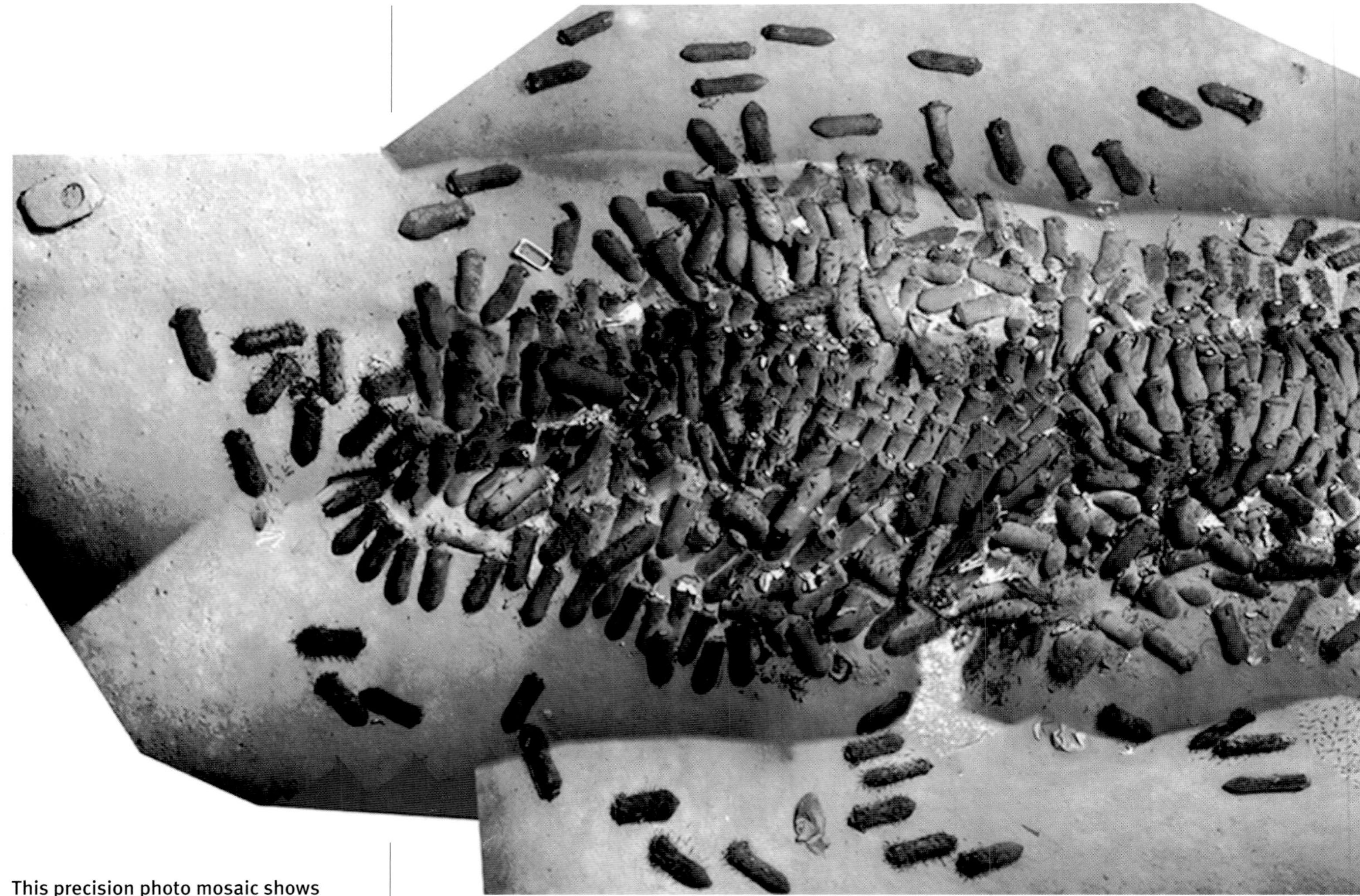

This precision photo mosaic shows the hundreds of intact amphorae once stacked in the hold of the Phoenician shipwreck we named *Tanit*. The stone anchor with a circular line hole lies at upper left; the two round cooking pots we retrieved lie at upper right.

suggested? Or did they fear the uncharted shoals of the North African coast just as today's seaman do? Unfortunately, finding a single Phoenician ship, however much we learned concerning its structure and cargo, would not reveal much about ancient trade routes, or the men who plied their wares along them.

The initial excitement of discovery began to give way to the discipline of underwater archaeology, a science that is amazingly similar to its counterpart on land. As on a terrestrial dig, we had to carefully survey and document the site before we could retrieve archaeological sample artifacts from this wreck and investigate the remaining *NR-1* target. But unlike above ground, we couldn't walk it off with graduate students pulling long tape measures and pegging down grid strings on the seafloor. Once more we had to rely on the versatile *Jason*. The ROV would produce an electronic photo-mosaic map of the wreck site.

To accurately map the wreck, we had to know precisely where *Jason* was at all times, so that the data from its digital sonar could be compared against known reference points. We met this challenge by deploying sophisticated navigation instruments on the ROV that communicated with a pair of electronic EXACT transponders anchored at either end of the wreck. The transponders acted as the equivalent of surveyors' transits, but using high-frequency sonar instead of light. Placing the transponders required a simple but elegant maneuver using an artifact-retrieval system we called the "elevator," developed and refined by Skip Gleason, another of my Wood's Hole colleagues: a green pipe-frame grid with yellow plastic floats on top and net compartments for the cargo equipped with sonar-activated controls to jettison ballast weight. Weighted by ballast, the elevator sank to the bottom, in this case carrying the EXACT transponders, and landed close to the wreck site. *Jason* approached and used its mechanical arm to unload and deploy the two transponders. The elevator would remain in place until artifacts were loaded into its net compartments. Then, on a sonar signal from the ship, ballast weights would detach, and the floats would lift the device to the surface.

Professor David Mindell, the navigational and electronics wizard from the Massachusetts Institute of Technology, took over from there. Once the survey navigation system was in place, *Jason* began the slow process of collecting its detailed electronic images, the precision sonar measuring the site's micro-topography as the vehicle hovered above the heaped amphorae, then creeping ahead to hover again. It was painstaking work, and many of the expedition gang, including Cathy Offinger, my former WHOI colleague and director of operations at the Institute for Exploration, pitched in to stand a watch, even though her schedule was more than full, managing the complex details of shipboard life during a scientific cruise.

By comparing hundreds of electronic close-ups with a smaller number of images taken at a higher altitude, the team was able to produce a precise computerized photo mosaic of the wreck site, using a four-foot wide CAD ink-jet printer that looked like a monster fax machine. Meanwhile, WHOI's Dana Yoerger and Louis Whitcomb from Johns Hopkins were hard at work on the digital bathymetric map of the site based on precision sonar data. This indicated our initial measurements were accurate.

The overall picture provided more clues to the ship's history. The false-color 3-D showed the amphora cargo lying in a neatly stacked heap about seven feet (two meters) above the floor of the oval depression, convincing evidence that the ship's hull had once been buried, then exposed, before being consumed by sea organisms. The orientation of the ship was east to west, with the bow pointing west, a possible hint that it had been heading toward Egypt or perhaps distant Carthage, if, in fact, the vessel had retained its approximate course after swamping. But we had gone as far as possible with speculation; it was time to get some answers.

Now that we knew so much about the ancient ship, it didn't seem right to keep calling it AC, so Larry's troops suggested the Phoenician name Tanit (pronounced Ta-NEET'). She was the Phoenician mother goddess who protected their seafarers and succeeded the Canaanite goddess Astarte. As consort to the important Phoenician deity Baal Hamon, Tanit was a stern mistress; parents sacrificed children to them both. But *Tanit* was clearly associated with seafaring, Larry told me: Several Phoenician monuments unearthed in Carthage carried the "sign of Tanit," a stick figure made with a circular head atop an isosceles triangle with upraised arms standing in the bow and the stern of stylized Phoenician ships just like the wreck below us. At the Phoenician level of the ancient port of Ashkelon, archaeologists had discovered bone and bronze amulets bearing the sign of Tanit, which were probably good luck charms for sailors. Unfortunately for the crew of our *Tanit*, their luck had deserted them during that horrible storm almost 30 centuries ago.

June 12 was one of those expedition days when cans of Coke and half a tuna sandwich between hurried conferences and catnaps punctuated the hours.

Then it was night again. *Jason* was back on board with the Woods Hole crew fine-tuning its systems, preparing for the demanding artifact-recovery phase on this wreck.

Larry and I had agreed on the fundamental principles for selecting artifacts. We would remove the minimum necessary to positively identify the wreck and provide useful archaeological information about the ship, its cargo, and its crew. And we would employ *Jason* to lift these materials free of the wreck in the least destructive manner possible.

Our first choice was to recover the cooking pots that lay on the open sand near the stern and a scattering of several amphorae on the edges of the denser stacks. But Larry and his archaeologists were still worried that *Jason*'s hydraulic arm could crush the terra-cotta artifacts, which might have become so cracked and waterlogged over the centuries that even the slightest touch would crumble them. I assured the archaeologists that *Jason*'s computerized hydraulic appendage could be adjusted to exert minimum clenching pressure, even less than a human hand. On this cruise, we were also using for the first time a new articulated grasping device that looked like two curved, oversize kitchen spatulas, one with open fingers. *Jason*'s crew had christened it "Deep Spank." In theory, a good pilot could use the hand set at minimum pressure to extract a reasonably heavy but fragile object from a narrow passage, while the robot arm was fully extended. Today would be a very good test.

Once more, the control van was crowded as the video screen and monitors cast their glow. Outside on deck it was a beautiful Mediterranean summer day. But within, only the clock revealed it was afternoon. I took my station beside Larry Stager, behind *Jason*'s pilot on this watch, Matt Heintz. The ROV approached the bulbous clay cooking pot, the floodlight casting a stark shadow against the pale sand. Matt worked the joystick like a laparoscopic surgeon performing a delicate procedure, slowly extending the arm, then gently thumbing buttons to fully open Deep Spank.

Larry was overwrought, and I couldn't blame him. The new hydraulic hand looked wicked, like something sprouting from a morphed character in a violent mutant cartoon. Twisting his own fingers to demonstrate how to handle the pot, Larry muttered, "Like that. Avoid the handles. They're not up to taking weight. I can tell you." This was like whistling past the graveyard. He had seen enough precious artifacts disintegrate on land where he had much more control of the situation. Now a robot was working down there in the icy darkness.

Matt slid the bottom fingers of Deep Spank under the pot and carefully closed the upper grip, then flexed the robot arm's wrist ever so gently. The pot rose slowly with a swirl of beige sand.

"That's the first time that's been moved in 27 hundred years," Larry said happily, the anxiety gone from his voice.

Archaeologist Dr. Lawrence E. Stager analyzes video and electronic images of the two Phoenician ships' cargo, a unique opportunity for him to work with such well-preserved artifacts.

Then the pot rolled free of Deep Spank and fell a few inches to the sandy bottom. We all gasped, waiting for it to shatter. The pot remained intact. Matt Heintz sank in his chair and shook his head. "Oh, man," he sighed in frustration.

Obviously, Deep Spank had not passed its first operational test. And I could see it was simply too dangerous to continue using the new hydraulic gripper with such fragile artifacts. We would have to go back to the original hand, two open opposing rectangular pipe frames strung with netting, which the crew called "the Cow Catcher." It didn't look as sophisticated as Deep Spank, but it had done the job for us when we first visited the Mediterranean in 1989.

"Okay," I announced. "We go to recover and change out."

Recovering *Jason* and changing the hydraulic hand would take precious hours away from time on the wreck. But we had no choice.

Once more *Jason* crept forward toward the curved flanks of the Phoenician cooking pot. Again, the tension in the control van was palpable. ROV pilot Martin Bowen, veteran of scores of such artifact retrievals using the robot arm and the Cow Catcher, seemed relaxed. He twisted the joystick and the robot wrist twisted on the video screen. Now the twin scoops of the Cow Catcher opened, the lower one easing under the flanks of the pot. With the lightest of button strokes, Martin commanded the upper scoop to close, then lifted the pot clear of the sand.

Larry Stager was beaming. "This is ideal."

Over the coming hours, Martin and the other pilots retrieved another cooking pot, a terra-cotta pot lid, and five amphorae, loading them aboard the elevator for two shuttle trips to the surface.

On the first elevator run from the bottom, we lined the starboard rail of the *Northern Horizon*, watching as the yellow floats broke the limpid blue surface. Crewmen in an inflatable Zodiac® were waiting to nudge the elevator beneath a ship's derrick and attach it to a winch line for lifting on board. Once more, Larry and his colleagues grew tense. The dripping green framework of the elevator swung back and forth, the darkly glazed cooking pot and a cylindrical amphora bulging precariously in the net, inanimate circus performers. If they fell, there would be no retrieving them, and I could see the Harvard crew was wary of the strength of the meshed lids on the artifact compartments.

Conservationists Dennis Piechota and Cathleen Giangrande inspect, clean, and tag material that our ROV *Jason* retrieved from the Phoenician wrecks off Ashkelon.

But we'd had a lot of experience bringing the elevator on board in all kinds of weather. Within minutes, it was safely lashed to the deck, and artifact conservator Dennis Piechota and his assistants were dutifully spraying the terra-cotta vessels with a mixture of fresh and salt water to keep them from drying too quickly in the hot afternoon sun.

Larry was almost dancing with joy. "Oh those beautiful cooking pots," he exclaimed. "So glorious!"

I leaned down to examine the dark terra-cotta pot with tight, dog-ear handles, which was filled with brown seafloor sediment. "What were they cooking in them?" I asked Larry.

His smile was even broader. "That's the kind they'd cook a one-pot stew in."

I joked that it was made for a Phoenician "refrigerator soup," into which the ancient sailors would have thrown everything from their refrigerator at the end of the week. But they'd had no refrigerators and depended instead on beans and lentils, stewed with fish, made savory with onions and garlic, and perfumed with marjoram and oregano their wives had picked and dried in their native hills in the spring.

I stooped and ran my fingers across the smoothly glazed rim of the pot. Had a stew been bubbling above a carefully tended charcoal fire in a sandbox when the storm struck? I knew Dennis's conservators would use the most modern scientific techniques to extract possible food residues from the pot. For me simply touching its surface evoked the heroic age of Homer. Perhaps I was the first human to have felt this cool, wet terra-cotta in almost 3,000 years.

When Larry's team moved the artifacts into their shipboard laboratory, he performed his initial close-up examination of an amphora. It lay in

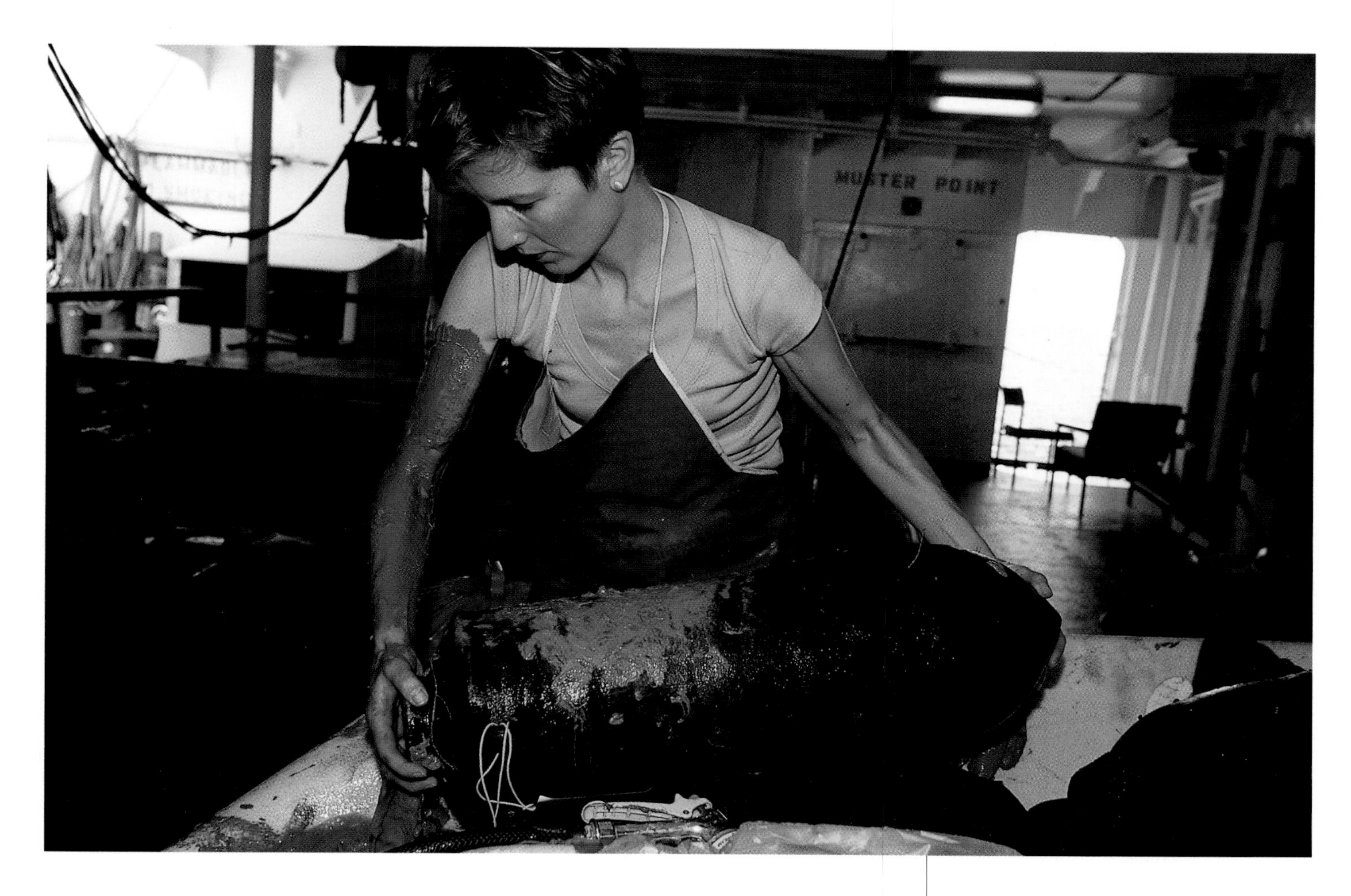

a white plastic cradle, dripping its preservative solution into a tub as Larry bent above. "Well," he concluded, "my date wish has come true. It's eighth century, not Byzantine." Larry added that on land he almost never found such pieces intact; they were invariably broken into sherds and had to be reassembled. But "out here," the deep sea was an impeccable museum, capable of preserving delicate artifacts from grave robbers, and natural disasters such as landslides and earthquakes. "There's a whole shipload of them intact," he added. "They are absolutely marvelous!"

Dennis Piechota and his son James worked with fellow conservator Cathleen Giangrande clearing the sediment from the artifacts. One of the principal questions about the amphorae was whether they had carried olive oil or wine. Both had been valuable cargoes in the ancient world, but the fine wine of Tyre and the Levant was especially prized. When Cathleen discovered traces of tree resin in one of the amphorae, we realized the cargo had been wine. Even well-glazed amphorae were somewhat porous. Olive oil sealed the pores. But wine seeped out unless the jars were coated inside with pitchy resin. Until the advent of wooden barrels and glass bottles, it's likely that all Mediterranean wine shipped and stored in amphorae had the piney tang of resin. Modern Greeks still have a taste for this retsina, which ranges in the fullness of its resin flavor from a subtle hint to what the uninitiated tourists call turpentine. Cathleen discovering resin traces in our Iron Age amphora confirmed the Phoenician ship had carried a precious cargo of wine.

Cathleen Giangrande carefully cleans one of the Phoenician amphorae. She discovered traces of resin coating one of the terra-cotta jugs we recovered, confirming the ship had carried a precious cargo of wine.

The expedition ship *Northern Horizon* recovers our workhorse ROV, *Jason*. The product of 25 years' research and development, *Jason* provides explorers aboard ship the precious element of telepresence on the hazardous ocean floor.

Even though we had finished artifact recovery at the *Tanit* site, we had to wait until first light on June 13 to recover the second elevator load of amphorae; the hours we had lost changing *Jason*'s hydraulic hand now caught up with us. We had to send the elevator back down for a third time so that *Jason* could retrieve the final four amphorae we had selected from the edges of the cargo heap.

As a result, we didn't begin towing *Jason*, now lifted a safe altitude off the bottom to avoid snags with uncharted obstacles, until well into the day. Even though we had a full week left on site, we were actually hard-pressed for time. Every hour counted. Not only did we have to locate and survey the next wreck as we had *Tanit* before beginning the complex task of recovering any artifacts, but also all the scientific work had to run parallel to the NATIONAL GEOGRAPHIC magazine and film crew coverage of the expedition, which was a vital but time-consuming aspect of modern exploration.

As luck would have it, several hours of *Jason* runs across the bottom failed to give us visual confirmation of the *NR-1* and the DSL 120's third sonar target. Maybe it had simply been geology after all. I had to make a choice about timing. We could continue to hunt with *Jason*, which was not a very efficient wide-area search vehicle, or we could go back to the DSL 120 side-scan sonar and make a last-ditch effort. But that would eat up more hours. The third alternative would be returning to the *Tanit* site for more filming and photography.

I was in the control van, just about ready to give up the hunt, when *Jason*'s short-range sonar began to paint a target dead ahead. Soon the wide video monitor revealed a chalky gray stack of amphorae in the glaring floodlight, and everybody was cheering again.

Larry Stager proclaimed the obvious, "It's another vessel."

As *Jason* rose to pass across the heaped amphorae, we saw the distinctive oval

shape of the cargo hold. "It's the same guy," I quipped, meaning that this ship and *Tanit* were practically twins, although the new vessel seemed larger. A greenish moray eel undulated through the amphorae, followed a moment later by a small stingray. As *Jason* reached the stern, we saw even more terra-cotta cookware and bowls strewn across the sand. "Looks like this ship found the same storm," I added. The two wrecks had obviously carried similar cargoes. Perhaps they had been part of the same convoy carrying wine to Egypt or farther west to Carthage. "Well," I said, "that wine company went bankrupt."

Our initial measurement showed that this ship was in fact larger, its oval mound 59 feet (18 meters) long. Larry's colleague Shelly Wachsmann proudly announced this made it the largest pre-Classical shipwreck ever discovered. A rough count revealed there were almost 400 amphorae in the cargo holds. This rich cargo had no doubt been cushioned by straw and matting and held tightly in place with a rigid frame, all of which had disappeared centuries earlier.

The Harvard archaeologists later named this vessel *Elissa* after the Tyrian princess and sister of Pygmalion, the legendary Phoenician king of Tyre and Cyprus who fell in love with an ivory statue of Aphrodite. Elissa had fled her homeland, sailed for the western Mediterranean, and founded Carthage by out-bargaining the natives in buying land for the colony. Myth held that she only desired territory the size of an ox hide: then she cut the hide into narrow strips and marked off a large area, including an acropolis known as the *Byras* (Fortress).

Scheduling pressure now grew a bit hectic, as so often happens on expeditions. We carefully logged the new site's coordinates, then *Northern Horizon* made the short transit to Ashkelon, where several personnel went ashore and others came aboard. We also welcomed Amir Drori, the Israeli Director of Antiquities, and his delegation. Larry Stager and I were proud to have *Tanit* and *Elissa* to show them.

By the afternoon of June 18, we were hard at work conducting our detailed electronic photo-mosaic and digital sonar survey map of *Elissa*. While the scientific effort was under way, the crew prepared for our educational obligations, rigging a down-looking 35-mm camera on our trustworthy old towed camera sled *Medea*, so that we could photograph the ROV during the artifact recovery.

Jason's cow catcher hydraulic hand proved to be the perfect device to scoop up the small clay cooking pots, shallow bowls, and an elegantly symmetrical chalice. The robot arm also safely retrieved a terra-cotta wine decanter with a slender neck and distended lips like the wide mouth of a flower.

I could see Larry Stager and Susan Cohen were gripped by anxiety as Tom Crook, a veteran of countless expeditions, nonchalantly tripped the switch in the control van to send the sonar signal that would dump the elevator's iron ballast weights and send the priceless load to the choppy late-afternoon surface. But once again, our experienced expedition crew gently swayed the elevator aboard and we had a chance to inspect our latest artifacts, even as the conservationists went to work with their plastic spray bottles.

We were surprised to find a well-worn stone mortar, possibly for the grinding of spices or incense—what had appeared to be a shallow bowl in *Jason*'s video. That fit with the next object, the chalice, shaped something like an uneven hourglass, about eight inches high. The wider, upper cup, Larry explained, was the incense chamber. "They'd burn bits of frankincense and myrrh in here over charcoal, evoking Ba'al." I wondered if *Elissa*'s crew had prayed to Ba'al and Tanit when the final storm had struck. Now Larry lifted the wine decanter into the clear Mediterranean sunset. "This is the clincher," he said, carefully fingering the artifact's delicate lips. "This is uniquely Phoenician."

The presence of these two Phoenician ships, so similar in design, bearing virtually identical cargoes, both resting on the bottom on an even keel, might have meant that they'd been sailing in convoy, perhaps the unlucky pair out of a much larger fleet. Or maybe *Tanit* and *Elissa* had actually sunk years or decades apart and had simply come to rest on the seafloor so close together by sheer coincidence; but speculation is inconclusive. At the end of the 20th century A.D., we have no way to determine what had happened at sea in any particular year 3,000 years earlier.

But our expedition would return bearing important knowledge about humanity's past. We had solved a mystery about Iron Age seafaring. Obviously, Phoenician merchant sailors had plied the deep sea, well beyond the reassuring sight of land. Discovering and exploring the wrecks to confirm their identity had established beyond doubt that the Phoenician voyagers had followed their explorer ancestors to venture forth beyond the ocean horizon—using the sea as a bridge, not a barrier—possibly to discover new lands, build long-distance trade routes, and plant colonies. Eventually risking extended voyages, these ancient seafarers bypassed laborious overland trade routes that were often blocked by difficult terrain or hostile peoples.

They had felt confident enough of their vessels and their voyaging skills to risk their lives and their immensely valuable cargoes by sailing this route. Although, the position of these two wrecks by itself did not reveal the ships' destination, the Harvard archaeologists were confident that, given the containers shape, the amphorae carried wine from Tyre in modern Lebanon. Such a cargo was probably destined for rich merchants in Egypt's Nile ports or in the Phoenician colony of Carthage, founded by the legendary Elissa herself.

As the expedition ended and the *Northern Horizon* headed toward Israel, I found myself thinking of ancient voyagers of the Mediterranean, "the Sea Between the Land," and of the courage it took for those nautical pioneers to explore its limits, then to venture forth beyond.

Three hundred years after *Tanit* and *Elissa* sank, perhaps the ancient world's boldest exploring seafarer, the Carthaginian Admiral Hanno, set forth on an epic voyage of discovery and colonization that has never been equaled.

During the fifth century B.C., Hanno assembled 60 vessels and 30,000 colonists, both men and women. His fleet sailed west from Carthage, out through the Pillars of Hercules—the Strait of Gibraltar—to the ancient Phoenician settlement of Tingis, present-day Tangier. Now Hanno and his captains faced the cold gray Atlantic. Undeterred by the alien coast, Hanno led his fleet farther south, founding fortress settlements along the coast. Today, remains of Carthaginian stoneworks cut by Hanno's masons can still be found in the Moroccan cities of Agadir and Essaouira. These are among the last of Hanno's ruins that can be verified, but it is probable that his fleet continued relentlessly south along the African coast, reaching the Gambia River and perhaps today's Sierra Leone. Significantly, it would have been there that they first encountered the most virulent tropical diseases and were obliged to return north.

Some scholars have disputed Hanno's voyage as mythical self-aggrandizement, mainly because an epic description of the accomplishment was inscribed in the Temple of Ba'al at Carthage and later transcribed into Byzantine Greek—at best a third-hand version of an actual explorer's log. And these scholars also find it hard to believe that Carthage could have built and crewed such a large fleet of seaworthy vessels.

After exploring the Iron Age *Tanit* and *Elissa* in the deep sea off Ashkelon, however, I have little doubt that 300 years later—with inevitable advances in shipbuilding—

Today one of the world's busiest shipping lanes, shown here from the Golden Horn, the Bosporus flooded the Black Sea basin in a cataclysm approximately 7,500 years ago.

Admiral Hanno did in fact venture forth on one of the longest voyages of exploration in the ancient world to return to Carthage with his precious cargo of knowledge of the huge Atlantic beyond the Pillars of Hercules. This knowledge allowed the Carthagenians to colonize the coast of present-day Morocco.

Successful exploration sometimes depends on mishap or serendipity. A voyager might be blown off course only to discover a new land, or the sudden outbreak of war might divert a scientific expedition.

This was the case in the summer of 1967, when geologists and chemists from the Woods Hole Oceanographic Institution were en route to the Red Sea aboard the research vessel *Atlantis II*. The Six Day War exploded, and expedition leaders David Ross and Egon Degen diverted the ship to the Black Sea to take bottom core samples and study the sea's unique hydrology.

The Black Sea is all but landlocked, with a single narrow outlet in the southwest, the Bosporus, which separates the European and Asian sides of sprawling Istanbul, then leads to the strategic chokepoint of the Dardanelles. Dense, salty water from the Aegean flows into the Black Sea through these constricted straits and sinks beneath the surface layer of fresh water constantly replenished by the discharge of major rivers draining Central and Eastern Europe: the Danube, the Dniester, the Dnieper, and the Don. This immense volume of fresh water creates an effect similar to that of a bathtub without a drain, the less dense fresh water overflowing through an escape valve (the Bosporus), trapping the heavier inflowing Aegean salt water in the depths for millennia.

Unable to circulate, the salt water in the Black Sea abyss has become completely depleted of dissolved oxygen. In this anoxic state, normal life is impossible. The lower reaches of the Black Sea are as sterile as an autoclave, an environment that exists nowhere else on Earth. Ross and Degen wanted to investigate these conditions.

But this was the height of the Cold War, and tensions were exacerbated by the decidedly hot conflict in the Middle East. The Woods Hole expedition had to stay well clear of the Soviet coast and the shores of its East Bloc satellites. But even on the open sea, a big Soviet Bear turboprop bomber repeatedly buzzed the ship at masthead level, tangible proof of the suspicion that precluded meaningful international scientific cooperation at this time.

The expedition was able to conduct a rigorous core-sampling survey of the deep-sea sediments. Gross examination of these core samples showed a layer of black mud sediment flecked with thin white strands above light gray clay. Under the microscope, the tiny white filaments proved to be the skeletal remains of marine plankton that had fallen as "snow" from the sunlit surface. But when the clay was processed in a press, it yielded water fresh enough to drink.

The scientists proposed a hypothesis: During the Ice Ages, the Black Sea became a freshwater lake as the world's oceans, including the Mediterranean, dropped dramatically. Then, when the glaciers melted, the east European rivers carried their immense loads of glacial clays into the huge lake where they were deposited on the bottom as the pale gray layer. Further warming brought a general rise in water levels, both in the Black Sea and the global ocean. Gradually the sea levels equalized, dense salt water flowed in through the Bosporus, and the present-day hydrology of the Black Sea took hold. The evidence for this argument was persuasive. The upper sediments were clearly marine. The lower clays not only contained concentrated vestigial fresh water, but also the remains of tiny freshwater fauna.

But Ross and Degen were not the only Americans to have studied the Black Sea. Six years before the fortunes of war took them there, a young Woods Hole graduate student named William B.F. Ryan first cruised to the Black Sea aboard the WHOI research ship *Chain*. Using a powerful new echo sounder, that expedition had discovered that the northern mouth of the Bosporus strait had slashed a deep chasm through the sediment and bedrock, an undersea river gorge that extended well offshore. The shape of this gorge and its relative lack of sedimentation suggested the feature had been carved with incredible speed and pressure. Bill Ryan went on to earn his Ph.D. in marine geology and become a senior scientist at Lamont-Doherty Earth Observatory of Columbia University. Deciphering the mystery of the Black Sea became his obsession.

He found some clues to the sea's intriguing past nine years later during a core-drilling expedition to the Mediterranean aboard *Glomar Challenger*. A deep-sea scientific and oil survey vessel, the ship extracted core samples from the bottom throughout the Mediterranean in the summer of 1970. In effect, the expedition was a more ambitious version of the *Atlantis II* cruise to the Black Sea in 1967 that sought to decipher the geological history of a sea by sampling its bottom. Studying the scores of core samples, Ryan concluded that the ancient Mediterranean had become separated from the planetary ocean approximately five million years ago, then dried to a salty desert. But the cores revealed a "razor thin" edge between the compacted desert sand and an unmistakable layer of marine ooze lying above. The ancient, desiccated Mediterranean basin had been inundated, Ryan concluded, in a single, cataclysmic event. But how?

He found his answer at the Strait of Gibraltar. Five million years ago, he theorized, that shallow strait had been dry land, separating the Atlantic from the Mediterranean basin. A violent seismic event had toppled this natural dam, causing the sea to flood through the narrow channel and fill the entire Mediterranean basin to an average depth of 5,000 feet in a monstrous deluge, the evidence of which was clearly delineated in the core samples.

For more than two decades, pioneering geologists Walter Pitman (left) and William Ryan (right) struggled to prove their audacious theory that the Black Sea was once a freshwater lake inundated in a single catastrophic flood. Our 1999 expedition helped confirm their hypothesis.

Ryan now joined his colleague Walter Pitman and British geologist John Dewey in a study of the great Eurasian mountain chain that began in Spain and continued almost unbroken into Iran. Their research on the mountainous region of the Black Sea suggested to Ryan that its basin might have once been completely cut off from the Mediterranean, just as that sea had been from the vastly larger Atlantic. Could the Black Sea also have evaporated to a vestigial lake, then been inundated in a single catastrophic event, Ryan wondered?

The three scientists speculated that such a cataclysm was a good candidate for the biblical account of Noah's Flood, but agreed it would have had to have occurred relatively recently in geological time because modern humans had only evolved in the past 100,000 years. For Flood legends to have permeated the cultures of the ancient Middle East as they had, the actual event would have needed witnesses who spoke coherent language, were grouped into society, and passed on legendary memory to the generations who followed.

Over the next 20 years, Ryan and Pitman developed a bold hypothesis, based partially on the intriguing *Atlantis II* expedition research and on Ryan's findings in the Mediterranean basin in 1970. They accepted that the level of the Black Sea had indeed dropped during the Pleistocene as vast ice sheets trapped so much of Earth's ocean water. But instead of the sea's basin remaining full from its river feeder system, the level in the now freshwater lake had slowly dropped and the shoreline retreated. Instead of an outlet, the Bosporus Strait became a riverbed draining down to the now-exposed northern grassland shore. Between about 12,500 B.C. when Ice Age melting commenced and about 5600 B.C., Ryan and Pitman believed, the freshwater lake remained isolated as global sea levels continued to rise.

Then, one day the sea lapped against the northern end of the Bosporus, cutting a tiny channel down the grassy slope. But within 60 days, the trickle had become a torrent, then an unimaginable cascade, gouging away rock and soil with a force two hundred times greater than Niagara Falls. This was the gorge Ryan had first observed profiled on the paper scroll of *Chain*'s depth sounder. Over the next two years, ten cubic miles of saltwater blasted through the newly formed Bosporus gorge every 24 hours to inundate the lake basin over 500 feet (152 meters) below. The water level in

This traditional view depicts the biblical Noah's Ark reaching dry land after the Great Flood. More likely, most Paleolithic people on the Black Sea shore escaped the flood on reed rafts, in dugout canoes, or simply ran for higher ground. But there would have been great carnage directly in the path of the Bosporus inundation.

the lake steadily rose to engulf over a mile-wide margin of dry land a day in areas. Eventually, the two geologists estimated, the level of the Black Sea rose 550 feet (168 meters) above the surface of the landlocked lake. None of the flora or fauna of this freshwater ecosystem could have survived. Pastoral people and Neolithic farmers, who had settled in the vast and bountiful oasis of the lake basin, drawn by its benign climate and rich soil, would have lost their grazing land, herds and flocks, and their fields. Uncounted thousands of these shore dwellers would have lost their lives.

This would have indeed been a catastrophe of biblical proportions, so devastating in the collective conscious that the legend of a Great Flood would have remained largely intact for hundreds of generations. Although it is improbable that a single patriarch such as Noah built a single immense ark in which to rescue birds and animals from the Flood, Ryan and Pitman's theory left ample room for Neolithic coastal dwellers—who most likely had reed-bundle rafts and dugout canoes—to

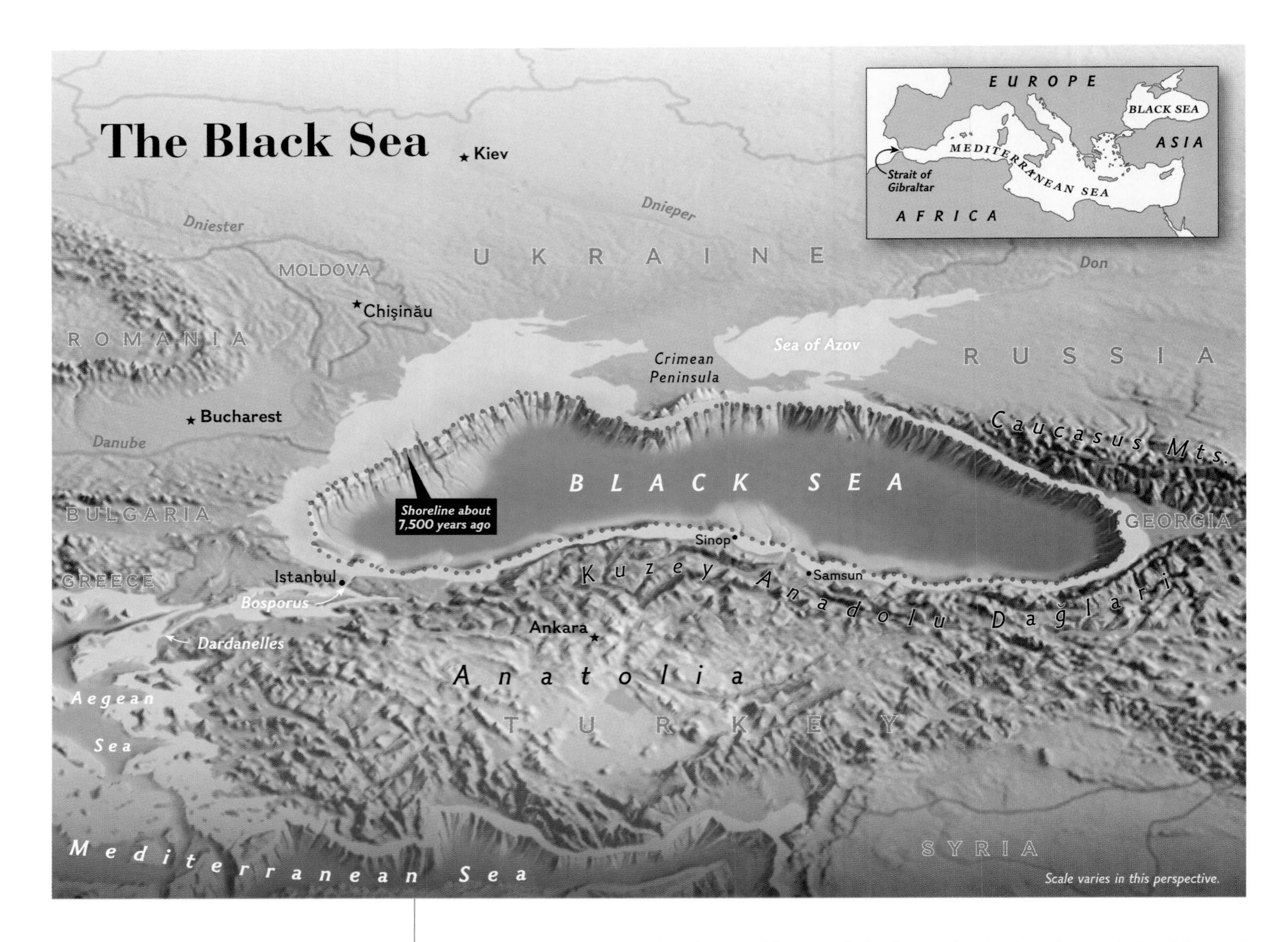

Ryan and Pitman's core samples revealed saltwater flooding along the Black Sea's northern coast. Our 1999 expedition found a well-preserved relic beach near the Turkish seaport of Sinop, which lies south of the Crimean Peninsula.

have rescued their families and livestock by boat, deeds that formed the fabric of heroic myth.

In fact, legends of a Great Flood can be found in many ancient cultures. Most notably, the Babylonian epic tale of Gilgamesh describes a wandering king who encountered victims of a Great Flood who had survived aboard an ark-like boat accompanied by clansmen and livestock. Similar myths had emerged from prehistory into European cultures. And the Spanish Conquistadors were astounded to hear Flood-survival legends from North American Indians. If Ryan and Pitman were right, the prevalence of these legends was logical: The northern hemisphere was the most heavily glaciated region during the last Ice Age, and the relatively rapid melting would have devastated the cultures of newly settled Neolithic farmers, an upheaval unlikely to have been forgotten.

Ryan and Pitman finally got the chance to test their daring theory in the summer of 1993, two years after the collapse of the Soviet Union. They unexpectedly heard from Professor Petko Dimitrov, a Bulgarian oceanographer, who had learned of their hypothesis and even tested it by diving off the present-day coast Danube delta in a manned submersible. At a depth of 404 feet (123 meters), Dimitrov found indications of an ancient beach. He had also collected shells from this beach, which he had subjected to radiocarbon dating that yielded a value of 9,000 years before the present. This suggested that the Black Sea had been a partially empty basin separated from the global ocean by some external barrier. In the 1960s, Soviet scientists had already found similar submerged landscape in the Sea of Azov, near the Crimean Peninsula, after engineers had taken core samples for bridges. But, at that time, the secrecy and suspicion of the Cold War had prevented international scientific cooperation, particularly between Russians and Americans, especially in such a sensitive area where the Soviets had major naval bases.

Now the Americans learned that scientists from the P.P. Shirshov Institute of Oceanography in Moscow wanted Western help in taking sediment samples near the Russian and Ukrainian coasts to track radioactive contamination from the 1986 Chernobyl' power plant disaster. Ryan and Pitman eagerly volunteered, offering to lend the expedition sophisticated new sonar equipment that could extend the survey offshore, and, in so doing, allow the Americans to search for ancient drowned shorelines. Candace Major, an undergraduate geology student, joined the two professors after cramming through a self-taught tutorial in fossil shell identification. She would prove to be an invaluable expert.

The Americans joined their Russian colleagues aboard the *Aquanaut*, a converted fishing trawler somewhat lacking in creature comforts, but still a solid seagoing vessel. Professor Kazimieras Shimkus, chief scientist, was intrigued by the Americans' hypothesis and by their sonar gear. He was willing to devote as much time as he could to the hunt for ancient Black Sea beaches before turning to the main order of business, the search for radioactive spread.

During the first sonar runs, the scientists found the clear profile of an ancient streambed buried under the bottom sediment. The sonar also revealed submerged outlet channels of the old Don River on the once-dry coastline prior to the Flood. Wherever the *Aquanaut* traced the bottom contours, the pattern was the same: Stream and riverbeds, now filled with river sediment, were visible in the sonar images. These waterways had once flowed out of the steppes to the north, across the grasslands of the exposed shore, and into the freshwater lake.

The Russians were very adept at core sampling; they used a cable-and-tube system that penetrated the bottom sediment to a good depth, returning a hefty sample. As Candace Major went to work on the compacted mud, she found tiny white marine clams, such as *Cardium edule*, in one layer of sediment, and brackish and freshwater shellfish in lower layers. Here was the evidence of a Flood Ryan and Pitman had been seeking for years: a marine environment quickly extinguishing an established freshwater ecosystem, with a very thin period of brackish transition.

In January 1994, Glenn Jones at WHOI's McLean Laboratory subjected samples of Ryan and Pitman's Black Sea shells to radiocarbon dating, using the highly advanced and accurate accelerator mass spectrometry technique. The dating revealed the shells, collected at widely dispersed sample sites, to all be between 7,580 and 7,470 years before the present. In other words, these marine species had migrated north in a great wave, not a gradual influx.

That would have placed the great Bosporus Flood well within the range of human memory. Agriculture and animal husbandry had been thriving for thousands of years in the Fertile Crescent and on the Anatolian plateau, not far from the shores of the Black Sea itself. Ryan and Pitman's research led them to believe that this cataclysm was not only the source of the Flood legend, but also the impetus for a general diaspora of prehistoric people out of the fertile Black Sea basin and into Europe and Asia. That was certainly an intriguing possibility.

The harbor of Sinop, here at sunset, provided a home base for our 1999 and 2000 expeditions. A port since the Bronze Age, ships have crisscrossed the Black Sea from Sinop to the Crimean Peninsula for millennia.

But I was much better equipped to test their hypothesis at the level of oceanography and geology than as an ethnographer. Again it was serendipity that influenced the course of exploration. I'd been working with colleagues for years to mount an expedition to the Black Sea, attracted by the unique and intriguing hydrography of the region. I had first read Willard Bascom's *Deep Water, Ancient Ships*, in which he theorized the Black Sea's anoxic depths would preserve ancient shipwrecks, 25 years ago.

Given the sea's sterile, anoxic bottom layers, any ancient shipwreck resting on the bottom might be perfectly preserved, unlike the *Tanit* and *Elissa*. And there might be plenty of ancient wrecks. The Black Sea had hosted a vigorous exploration and maritime commerce beginning in the Bronze Age (c. 3500 B.C.) or perhaps even earlier. It was only logical that a large number of ancient ships, ranging from war galleys to heavily laden merchant vessels, would have foundered and sunk to the lifeless black depths, to rest undisturbed on a bottom devoid of the wood-boring worms and smaller organisms that have consumed the planks, rigging, and sails of every ancient shipwreck yet discovered on a normally oxygenated seafloor.

So, supported by the National Geographic Society and the J.M. Kaplan Fund, I assembled a team from the Institute for Exploration and invited sonar expert Dave Mindell of the Massachusetts Institute of Technology and archaeologists Fred Hiebert of the University of Pennsylvania and Cheryl Ward of Texas A&M University to direct the principal land and undersea surveys. We met in Turkey in the first week of July 1999, just a month after our successful expedition off Ashkelon that had discovered the Phoenician ships *Tanit* and *Elissa*. Our original goal was to find evidence of a well-preserved ancient port on the coast and shipwrecks just offshore in shallow water, then broaden the search by moving north into greater depths, hunting for a trail of "mint-condition" wrecks that would mark the trade route between ancient Sinop (on the north coast of present-day Turkey) and the Crimean Peninsula, of which Classical historians had written.

While our expedition was in its final planning stages, however, we were astounded to read Ryan and Pitman's research findings and were also engrossed by their compelling popular account, *Noah's Flood, The New Scientific Discoveries About the Event That Changed History*. They had devoted much of their professional lives to the Black Sea Flood hypothesis. Their long effort deserved support. In science, research findings only become valid if they can be replicated. Ryan and Pitman had found evidence of a sudden inundation of a freshwater lake in today's Black Sea basin on the ancient sea's northern coast. Their most compelling discoveries had been apparent submerged beach formations at an average depth of 554 feet (169 meters) below present sea level and species of marine shellfish dating from 7,500 ago. Could my expedition replicate their findings on the south coast, thus completing the picture of a flood of true Biblical proportions engulfing the entire Black Sea in a single cataclysmic event?

Bracing my legs against the chop, I gazed at the amber swirls creeping across the computerized monitor of the side-scan sonar. The little Turkish trawler *Guven* plowed slowly through the sunny Black Sea. Below us, the torpedo-shaped sonar glided on its cable tether near the gravely bottom at a depth of 550 feet (168 meters), sending up a digitized acoustic image. I felt the boat's engine rumble as we completed this northerly survey line. The captain up in the tiny bridge above this improvised laboratory had spun his wheel hard over and the wooden-hulled vessel swung back to the south so that we could acquire a sonar picture that overlapped the previous "transect" line on a new, reciprocal heading.

We launch the small ROV *SeaROVER* during our Black Sea 1999 expedition. Although we had high-tech equipment, we eventually resorted to a simple dredge to retrieve shell samples for radiocarbon dating that helped prove the theory of the Black Sea's Great Flood 7,500 years ago.

The electronic equipment stacked on the plank tables in this cramped fo'c'sle represented the state-of-the-art in sensitivity and miniaturization. But the goal of the multi-discipline, international expedition I was now leading was one of the most exotic I had ever been involved in during almost four decades of sometimes-bizarre endeavors. I had to remind myself that we had come to this stretch of water, twenty miles east of the Turkish town of Sinop, site of an ancient Greek port and one of the few natural harbors on the southern Black Sea coast, to determine whether the biblical account of Noah's flood—as related in the Book of Genesis and in similar accounts in other cultures—was simply poetic myth or verifiable scientific fact.

From their 1993 cruise on the *Aquanaut*, Ryan and Pitman believed that the ancient lake's original shoreline should be near the 550-foot bottom contour of the present-day Black Sea. At least their sonar surveys and core sampling had so indicated off the Russian and Ukrainian coasts.

But as I studied the big Black Sea hydrographic chart, I tried to imagine the actual cataclysm of the Great Bosporus Flood. What would that disaster have been like to the Neolithic people living in scattered villages of clay-and-wattle huts on the grassy hillsides above the lake when the upland valley burst above them? The relentless force of the global ocean, almost three-quarters of the planet's surface, would have blasted out that narrow gorge in a monstrous, deafening torrent of green salt water and towering spray. I thought of Hoover Dam bursting at its base, then multiplied the disaster a thousand-fold.

The people living in the path of the Flood would have been instantly killed; their villages and herds swept away like chaff. The full blast of the Flood would have eradicated the ancient lakeshore features in its path and dumped masking sediments over a wide area of newly submerged sea bottom.

But I wanted to find convincing evidence of the now flooded ancient beach that had neither been destroyed by the Flood's violence nor buried under the river sediments that continue to pour into the Black Sea basin long after the Flood.

So I knew I had to stay away from both the Bosporus and the European rivers dumping into the sea's northern coast. I wanted to get "downstream" of the Flood, as one would on the Mississippi River at New Orleans in the spring. When violent floods ravage the Illinois levees, water creeps up by inches in the Big Easy. Could I find a similar area where, 7,500 years ago, the water of the ancient lake simply began to rise slowly in response to the Flood, inundating the ancient shoreline mere inches each day with-

In the cramped quarters of a small Turkish trawler, *Guven*, Martin Bowen (left) drives his ROV *SeaROVER*, while Sarah Webster (center) and Brendan Foley (right) observe. Dr. Dave Mindell and I (background) peek through the light-obscuring screen.

out destroying its fragile beach features or burying them beneath a blanket of sediments? And could I discover this ancient preserved shoreline exposed beneath the modern Black Sea at the precise depth Ryan and Pitman predicted I could find it?

I thought that I could. I decided the region off the Turkish coast near Sinop might be an ideal place to conduct my search. It was many miles east of the Bosporus, and well away from any sediment-bearing rivers. The Flood waters, I believed, would have risen gradually, preserving the distinctive contours of the original lake's beach.

Had I known investigating the Black Sea Flood theory would become one of the expedition's major goals, however, I would have tried to have a larger, better equipped research vessel, as well as a flotilla of camera and video sleds and our old faithful ROV *Jason*, which, unfortunately, was currently unavailable. But, if Ryan and Pitman's bold theory was right, I was confident we could verify it relatively quickly with the equipment and vessels we did have.

The composition of lake coastlines is familiar to every professional geologist, whether the shore be on Lake Erie or the Caspian Sea (another ancient landlocked lake). I had spent long nights as an undergrad studying these formations. On the windward coast, prevailing wave action sculpts the bottom into typical forms, which include a beach of mixed cobblestones and smooth oval pebbles, a slope of sand and shells where the waves begin to drag bottom and break, a scalloped, hummocky inshore lake bottom, and often a humped sandbar formed by strong currents running along the shore. On the farthest landward side, there is usually a beach berm, the typical high-water line of eroding waves, then farther from the lake, a scarp or bluff.

I sort through material that we dredged from the submerged beach 550 feet below the present level of the Black Sea during our 1999 expedition. Radiocarbon dating of the fresh- and saltwater shells we recovered provided conclusive evidence of Ryan and Pitman's ancient flood.

If our expedition could capture sonar and video images of such an unmistakable lake shoreline—now submerged below the surface of the Black Sea—and also retrieve freshwater shells and wave-polished pebbles from the appropriate contour section, we would have made a major contribution to the three decades of dogged investigation Ryan and Pitman had already completed. As on any expedition, however, time on the site is expensive and precious, and conditions are rarely perfect. With the prevailing summer northerlies kicking up a nasty daily chop off Sinop, our small fleet of chartered Turkish trawlers was often unable to hold exact station over a particular stretch of bottom the way a more sophisticated research vessel equipped with dynamic thrusters linked to the global positioning satellite navigation system, such as *Northern Horizon* does routinely. We did have GPS to plot our heading, but no side thrusters to keep station over a particular piece of bottom.

So I knew the towed side-scan sonar—which was less affected by rough conditions than our two small ROVs, the *SeaROVER* and *Benthos-Mini Rover Mark II*, equipped for sonar and video imaging—might prove to be the most important exploration instrument at this phase. I had asked the *Guven's* captain to head almost due east of Sinop at full speed today until we intersected the 510-foot (155- meter) depth contour that meandered generally northwest to southeast, paralleling the foothills of the coast, 20 miles inshore. We then began our sonar survey tracklines, cutting in narrow, zigzag patterns, north and south over the bottom, out to a depth of 600 feet.

Now we were headed south again, running parallel to the previous trackline and approaching some interesting patterns of bottom structure at 550 feet (168 meters), very near the depth that Ryan and Pitman had predicted the ancient shoreline should lie. Arnie Carr and Rob Morris of the American Underwater Search and Survey Company hunched over their equipment to adjust the sensitivity of the side-scan sonar. As the tan image dissolved from the shadowy flecks of the deeper lake bottom into the smooth sandy surface, my eye shot to the geometric profiling model coupled to the computerized depth sounder. Just as on the previous northbound trackline, we appeared to be crossing the hump of a sandbar, bound for the ancient beach. Now the bottom dropped away again to resume the rippled pattern we had recorded on the last northbound survey line. The sonar image smoothed once more as the depth decreased and our sonar glided up a slope of smooth small detritus, most probably mixed sand and shells.

Suddenly this slope steepened sharply into the profile of the beach berm.

"Look at that," Cathy Offinger said, her voice edged with excitement.

We were crossing an almost flat cobblestone shore, and then a low bluff before reaching a bottom that stretched undifferentiated into a featureless coastal plain extending all the way into the present Black Sea coast of Turkey.

The people around me in the makeshift control room tried to suppress their fervor. But Cathy broke into a broad grin. "Yes," I said, "I think we might have nailed it." On the next survey line, the distinct curve of an ancient lake coast was even more apparent, again at a depth of 550 feet, only 40 feet deeper than Ryan and Pitman's original estimate. Yet we were hundreds of miles south of the survey lines they had run in 1993. I believed we were looking at the beach of the ancient freshwater lake, just as the two Columbia geologists had predicted.

The crisp images that the side-scan sonar had transmitted up the coaxial cable seemed indisputable: A submerged shoreline existed virtually undisturbed 550 feet below the throbbing wooden hull of the *Guven*. Logic mandated that the only way the scarp, cobblestones, berm, pebble bands, and sandbar could have formed in that exact sequence was on an ancient shoreline that had later been submerged by a flood.

My strategy in coming out to this section of offshore sea bottom seemed to have paid off. Ryan and Pitman had used sonar and core samples to search for their "relic" lake shore off the Crimea, where the immense deposits of sediment from European rivers would have long-ago buried the kind of exquisitely fine details we had just traced by sonar. The unimaginably powerful blast of flooding seawater would have carried millions of tons of masking sediment in all directions from the newly gouged Bosporus. But there was a basic tenet of geology that floods—and sediment-bearing rivers—dumped their loads exponentially. I had to search for the ancient lakeshore at least 200 miles from the Bosporus and off a coast free of rivers. Behind Cape Ince, the land rose steeply into knife-edge mountain ridges; there were no muddy rivers. If the relic shore existed, it might well be here.

As the little trawler headed back toward the steep beige-and-green slopes of Anatolia, I was reasonably confident we had just discovered proof of the Biblical Flood and felt the churning joy of successful exploration.

Still, I was a scientist; it was my professional duty to consider opposing hypotheses. Ryan and Pitman's long work necessitated such rigor. It was possible that our side-scan sonar had traced some previously undetected phenomena associated with the contact between the deeper anoxic waters and the well-oxygenated upper levels of the Black Sea. Another possibility was that, by incredible coincidence, we had discovered the "relic" shore of a lake that had formed in the far distant past, maybe hundreds of thousands of years ago during one of the periodic warming cycles of the Pleistocene Epoch, which had lasted over a million years. Finding that type of coastline would not confirm Ryan and Pitman's recent Bosporus Flood hypothesis. The best way to be certain we had discovered the Great Flood beach was to turn loose our capable little ROVs, equipped with precision video and still cameras, which might capture images of extinct freshwater shells.

Pioneer undersea archaeologist George Bass explored a Bronze Age wreck in relatively shallow water on the Turkish coast in the 1960s. Bass visited our 1999 Black Sea expedition.

I worked late that night with the ROV crews, reshuffling schedules and dividing our limited resources between the Turkish chartered vessels to maximize our effort. But a frustrating week followed our initial success. Strong northerly winds rose almost each morning and kicked up a bad swell by afternoon. Despite the best efforts of the experienced Turkish trawler skippers and our own veteran ROV crews, we were unable to keep the high-tech unmanned submersibles precisely on position for very long in these choppy, relatively shallow waters. In several quick "bounce dives," the *SeaROVER* did get some intriguing pictures of what was undoubtedly the beach berm running along at a depth of 527 feet (161 meters), then a low sandy terrace farther up the slope. On the submerged beach itself, our ROV operator Martin Bowen skillfully drove the *SeaROVER* into the sediment to dislodge a collection of scattered white rocks of a type not found near today's Black Sea coast. When the ROV was hoisted back on board, the recovery crew found a small black basalt pebble and white flakes that could have been freshwater seashell.

That was one of the few successful high-tech surveys. The gusty winds and harsh swells continued to plague us, making it impossible to "fly" ROVs steadily across the relatively shallow bottom from the trawlers that lacked the GPS-linked thruster systems of sophisticated ships such as *Northern Horizon*, which could hold the surface vessel precisely on station. High-tech was not available; I chose a decidedly low-tech option. Captain Idin Yildiz, whose family owned the steel-hulled trawler that we had dubbed the "*Z Boat*" because it hailed from the Black Sea port of Zonguldak, came to our rescue. He retrieved a shellfish dredge from weeds at the Sinop dock, and had his crewmen weave a purse-pouch behind the rusty iron fingers.

Once out on the site of our sonar-surveyed lakeshore on July 17, the captain's sailors simply dumped the dredge overboard when we signaled them that the GPS indicated we should be approaching the slope of pebbles, sand, and shells at a depth of 478 feet (146 meters). We then towed the heavy iron dredge out to 491 feet (150 meters), and then the boat's creaking winch hoisted it on board. As we all clustered around its dripping muddy purse, I probed the clotted mud with my fingers. My eye went to clumps of dark pebbles layered like uneven stacks of poker chips. The edges of the uppermost pebbles were coated a rusty orange where they had been exposed to the sea.

The rusty smudge was phosphorite and manganese that had precipitated out of the water over millennia to form a natural cement. Geologists recognized that this process occurred when a surface was flooded gently with sediment-free water. We had more proof that the relic beach below—far from the destructive muddy gout of flooding Bosporus—had been inundated gradually, probably just a few inches each day, so that delicate details such as the sculpted sand ripples and the low berm had been preserved intact. These pebbles, once incessantly polished by waves on the leeward shore of the ancient freshwater lake, had been submerged, and so shielded from breakers for millennia.

I had invited pioneering underwater archaeologist George Bass, who in the 1970s had dived on one of the first ancient wrecks ever in the Aegean—the Minoan-era ship he discovered on the Turkish coast—to join the day's work. I broke off a smooth, grayish pebble and tossed it to George. George wiped the pebble on his shorts and held it to his eye, noting the wave-polished texture of the stone. On the windward coasts of many Aegean islands, the local people have collected such *kouklakia* pebbles for centuries and set them in ornate mosaic floors in their houses and churches.

"I sure did come out here on the right day," George said.

Repeated dredges brought up shells and sodden chunks of wood. Over the next two days, we zigzagged back and forth across the submerged shoreline, dragging Captain Yildiz's primitive but effective dredge. Sometimes we came up empty. But on several hoists the purse was loaded with brown or white shells. We had retrieved the precious biological samples we needed for radiocarbon dating. As the *Z Boat* returned to Sinop on the last day, I knew these shells represented our best hope of establishing beyond reasonable doubt that Ryan and Pitman's catastrophic flood had occurred. If any of the shells were from freshwater species and if radiocarbon dating gave them an age greater than 7,600 years, our expedition would have confirmed the Great Flood.

The wave-rounded pebbles we dredged from the relic beach near Sinop showed the unmistakable polishing they received over thousands of years on the shore of the ancient freshwater lake.

On November 17, 1999, my colleagues and I met the world's press at the headquarters of the National Geographic Society to present the findings of our first Black Sea expedition. Our side-scan sonar profiles of the submerged shoreline of the ancient freshwater lake offered compelling evidence that Ryan and Pitman's bold thesis was indeed accurate. But the most persuasive confirmation came from the shells dredged up from the ancient beach. Gary Rosenberg of the Academy of Natural Sciences in Philadelphia had positively identified seven species of saltwater shellfish among our dredge samples, including *Mytilus* and *Trophonopsis*. But he had also found two extinct freshwater species—*Turricaspia* and *Dreissena*—similar to those living today in the fresh water of the Caspian Sea. Samples of each species had been sent to the Woods Hole Oceanographic Institution for radiocarbon dating. These tests showed that the saltwater shells had lived from about 2,800 to 6,820 years ago.

But, most dramatically, the freshwater species predated the others, living in a lake from 7,460 to 15,500 years ago. They had become extinct around 5400 B.C., when the deluge of salt water had killed them, almost exactly the time Ryan and Pitman had theorized the Great Flood. The extinction date of these freshwater species coincided with the arrival of the marine shellfish species that Ryan and Pitman had collected on the Black Sea's northern coast in 1993. We had closed the circle. No one could dispute that a Great Flood had occurred approximately 7,500 years ago.

II The Mysterious Sea

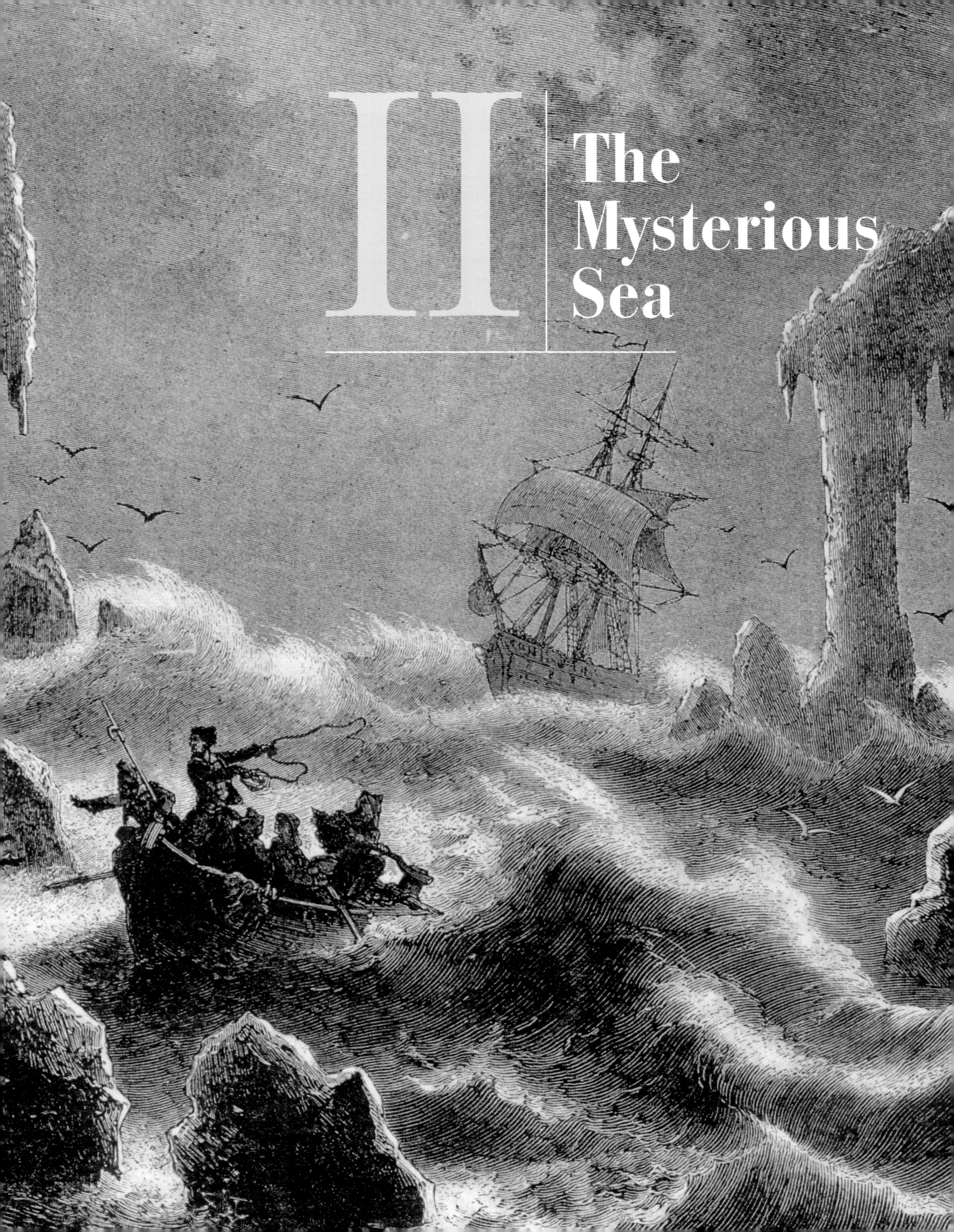

We carefully unload bottom samples that the submersible *Alvin* had retrieved from the Mid-Atlantic Ridge more than 9,000 feet below during Project FAMOUS, 1974.

became heavier than water; then the vessel sank at a maximum speed of a hundred feet a minute. To stop descending and achieve neutral buoyancy, the pilot released steel shot ballast stored in magnetized silos. Horizontal movement and short vertical climbs along the bottom were achieved with small electric thruster propellers. To ascend to the surface, the pilot jettisoned more shot, and the buoyancy of the gasoline float lifted the craft.

I swallowed with pain as I climbed down the narrow access tube, having woken in the night aboard the ship with a raw strep throat and fever chills. But I was stubbornly determined to make this dive. It represented that intangible point where science and exploration intersected at the most engrossing level. The day before, my French colleagues Xavier Le Pichon and Jean-Louis Michel (with whom I would discover the wreck of the *Titanic* in 1985) had made the first dive on the four-mile-wide canyon between the twin ridges. They had returned with lava samples and described traversing steep lava cliffs that seemed to confirm our view that this part of the Ridge was a fracture zone at which the North American and African plates were pulling apart.

Even though the clumsy bathyscaphe was not an ideal for such exploration, when compared to the nimble little American submersible *Alvin* or the French *Cyana*, which were scheduled to dive on the Ridge the next year, I wanted to be part of the team that initially explored those mysterious deep-sea mountains and canyons and brought back the first priceless scientific knowledge.

Inside the confines of the sphere, the wiry, taciturn engineer I knew simply as Semac showed me the observer's station at the aft end. Among its other limitations, the bathyscaphe had only one tiny viewport, which crew members used via connected binocular optics. This was a prudent precaution against the catastrophic pressures deep-diving submersibles confronted. But such restricted viewing made work difficult for a hands-on geologist and slightly dampened my explorer's zeal for observing part of the planet no human eyes had ever before seen.

Once Harismendy and Semac had meticulously verified the seal of the access hatch, they read a slow litany from a printed predive checklist; I couldn't follow their French. The pilot and engineer toggled switches, and I heard a *whoosh* as gasoline vented from the float tank. We were on our way to the bottom, almost two miles below.

Our dive site was on the eastern side of the Ridge, a series of steplike vertical cliffs, or "scarps" in geological jargon, that Xavier and I believed would be split by smaller side fissures and offer a good selection of fresh lava flows from which I hoped to retrieve samples. The steel-walled sphere quickly became clammy as we plunged deeper, down below the sunlit upper layers of the sea, into the eternal night of the abyss. I huddled in the aft end sheltering my swollen throat with my wool turtleneck. Had I felt better, I would have watched through the viewport optics during the 90-minute descent, trying to catch glimpses of bizarre creatures such as the tiny, but fierce looking hatchetfish, which had pulsing, neon-hued light organs along its belly. Below about 2,000 feet, the phenomenon of bioluminescence vanished. The scattered free-swimming or bottom-dwelling animals that lived in the true depths existed in a near-freezing world utterly devoid of light. To navigate to our survey site, Harismendy relied on a network of three acoustical "lighthouses" that my colleagues at Woods Hole had invented. Sending sonar pings of a specific frequency, Harismendy interrogated the buoys, which were submerged on anchors near the bottom; the buoys responded, and he plotted our position using modified marine triangulation.

I dozed feverishly while we dropped, only to snap awake when I heard a chirping alarm from the pilot's control panel. Harismendy switched on two powerful floodlights then tripped a switch to dump some ballast shot and raced the downthrusting prop. We hit the rocky surface with a jolt that boomed dully inside the steel sphere.

"On the bottom, Bob," Harismendy noted.

I slid over to my eyepiece and saw a swirling cloud of dirty beige sediment kicked up by the propeller. Few people realize that this grayish mud covers most of Earth's surface and is composed of marine "snow"—the biological detritus of the trillions of tons of microscopic, invertebrate, and vertebrate organisms that live and die in the oceans' upper layers; their bodies and digestive waste have drifted to the seafloor over countless millennia. Along the mid-ocean rift, where the seafloor is young, the sediment barely forms a thin dusting. But near the continental edges, where the ocean floor formed more than 150 million years ago, the layer of sediment can be thousands of feet thick.

Harismendy maneuvered the bulky submersible with amazing deftness as we descended the black staircase of the scarp face. The personnel sphere twanged as we jolted to a halt on the rocky bottom. This time through my eyepieces I could clearly see heaps of mounded black lava, whose gleaming surfaces reminded me of burnt

A mosaic shows a fissure, *gja*, cutting across the floor of the rift valley of the Mid-Atlantic Ridge.

bread loaves. There was almost no sediment on this lava; it was fresh. Molten lava at 2,200° F extruded through a fissure fought a brief, one-sided struggle against the near-freezing cold and crushing pressure of the seafloor. At this depth, 9,200 feet, the ambient water pressure was 300 atmospheres, just over two tons per square inch. Small "squirts" of lava were quickly quashed. Only major fracturing caused by earthquakes allowed enough molten material to escape from the mantle to heap up these steplike parallel ridges. I felt the tingle of discovery, seeing these gleaming black lava tubes, which were directly connected to Earth's churning mantle of superhot magma. Being one of the first humans to witness this previously unknown process of creation sparked in me a thirst for exploration that I have never slaked.

As Harismendy grappled a nice chunk of lava with the bathyscaphe's awkward mechanical claw, I pressed my face to the eyepieces, now grateful that this old clunker of a submersible had just one small, armored-glass viewport. Even the tiniest leak at this depth would unleash a rapier of pressurized water that would slash the three of to bloody pieces in seconds.

Harismendy bounced us down to another sample site, which I tried to designate using a mixture of English and my pidgin French. He got the gist of my directions, but every time he tried to snag an intriguing sample of lava that had solidified into a shiny black tube, the bathyscaphe tilted away and he had to rev the thruster prop, which clouded the floodlit field of view with light sediment. As we were maneuvering to get a better angle, I heard Semac, the engineer, mutter, "*Ah, merde.*" He tapped a meter on the panel. A large needle had dipped ominously into the red zone; we had just been hit by a major power failure. Then the bathyscaphe suddenly plunged bow-down, and I almost lurched over the two crewmen. Managing a four-limb perch like a monkey on a branch, I saw the glowing red digits on the depth sounder blinking too fast to follow. The power failure had cut off the electromagnets in the ballast silos and we had just dumped all the steel shot on board.

Harismendy turned to face me in the dim glow of the instrument lights and grinned apologetically. "Going up, Bob."

So be it, I thought. The scheduled dive had been almost over. We'd retrieved a nice lava sample. And I certainly could confirm the staircase structure of the east Ridge face. All in all, not a bad geological field trip. Beyond that, I was only the second scientist to actually view the Ridge at close hand.

My stomach grumbled pleasantly as I watched Harismendy open the plastic wrap of our bread-and-cheese lunch, while Semac pulled the cork from the bottle of Beaujolais we'd sipped on the descent. The French, I thought, were a very civilized people to go exploring with.

Then Semac jammed the cork back in the bottle and gripped Harismendy's shoulder, hissing the word *incendie*. Suddenly I smelled the frightening stench of burning electric wires. The two crewmen worked in a frenzy, cutting power from the submersible's battery bank. But their efforts were fruitless: Choking smoke had already filled the tiny compartment. My eyes flooded with stinging tears and I wheezed painfully. What a hell of a place to burn to death, more than a mile beneath the calm sunlit surface of the Atlantic.

Harismendy and Semac dragged out their emergency oxygen masks and I seized mine, trying desperately to remember the instructions in the system's use I'd been given. My throat was almost closed with hot spasms as I pulled the full-face mask over my head, tightened the elastic straps, and tried to breathe. Instead of the soothing flow of cool oxygen I expected, I choked down a lungful of acrid smoke. In a panic, I ripped the cloying rubber mask from my face. But Harismendy and Semac, like goggle-eyed monsters with elephant snouts, seized me and pulled the mask back down. Again I tried to breathe, only to choke on smoke. Finally, Harismendy lurched past my neck. "*Pardon*, Bob, *pardon*." He spun open a valve on my emergency oxygen panel, which we'd forgotten.

"Now eets all okay, Bob," he soothed.

I clung to the rear of the compartment, sucking deeply as the oxygen hissed through the mask, soothing my burning throat and drying my eyes. The thudding pulse slowly eased in my temples. After a long time, I saw a faint blue glow in the optics of my eyepieces. We were nearing the surface. I would live to explore another day.

In July 1769, a flat-bottomed converted North Sea coal hauler named *Endeavour* pounded into a Southern Ocean gale, the unwieldy ship making slow progress under close-reefed sails. In the rigging, miserable sailors crawled on the yardarms flailing with knotted ropes to break loose rime ice. On the sharply tilted quarterdeck stern, the vessel's commander, British Royal Navy Lieutenant James Cook, stood wrapped in a sodden oilskin cloak, watching the overcast sky hopefully for a break that might allow him a sextant sight of the sun. The task proved hopeless: At 40° south latitude, the full strength of the Southern Hemisphere winter had exerted absolute dominance of the sea.

Captain James Cook completed his first circumnavigation between 1769 and 1771 aboard this former North Sea coal hauler, *Endeavour*. The vessel's deep holds gave ample room for stores during the long voyage. Its double-planked hull and flat bottom made it ideal for treacherous seas and charting shallow waters.

Cook and his brave crew could venture no further south that season in search of their elusive goal, *Terra Australis incognita*, the mythical great Southern Continent that philosophers from Medieval times to the Enlightenment had believed must exist to "balance" the known northern landmasses.

I've always found it fascinating that these mythical continents were thought to be well watered and fertile. The missing Southern Continent was also considered huge: The famous 1586 global map of Abraham Ortelius showed the continent spanning the entire bottom of the Earth, and reaching as far north as the Tropic of Capricorn in several places. Sighting the mysterious Southern Continent and claiming it for the Crown were the principal goals of this voyage, Cook's first major attempt at exploration.

Both the Admiralty and the Royal Society believed that, if anyone could find the mysterious Great Southern Continent, it would be Cook.

After enduring incredible hardships in the Southern Ocean winter, Cook dutifully steered west, fighting the incessant gales, but following his Admiralty instructions until he encountered the North Island of New Zealand that October. For the next six months, Cook and *Endeavour* explored and charted the entire coastline of both the North and South Islands. They then explored the east coast of Australia from Botany Bay north through the perilous coral-shoal waters of the Great Barrier Reef.

On return to England in July 1771, he was received by King George III, the Admiralty promoted him to commander, and the Royal Society recommended that he

lead another expedition to the South Seas to prove or disprove beyond a doubt whether the mysterious Southern Continent existed. Cook now commanded a two-ship squadron of much more seaworthy vessels, *Resolution* and *Adventure*, both modified new North Sea colliers.

Cook meticulously prepared for his voyage before setting forth. On the previous expedition, he'd had no certain means of determining his longitude (the distance east or west of the prime meridian); in order to search for land, he had been

obliged to sail along a parallel of latitude determined by sextant sightings. But on this voyage, Cook had equipped his ships with a new instrument that would forever solve the problem of longitude: John Harrison's chronometer.

Harrison's marine chronometer was driven by complex springs, gears, and counterweights. It kept time much more accurately than pendulum clocks, which were of little use in the rolling, pitching cabin of a ship at sea. The principle of the chronometer was that the planet rotated 360 degrees every 24 hours, or 15 degrees each hour. A chronometer set at noon in Greenwich near London, the Prime Meridian, would retain that time. By comparing Greenwich chronometer time with "local noon," as determined by a sextant sight of the sun at its zenith, a navigator could accurately assess how many degrees east or west of Greenwich he was: the ship's longitude. Now Cook, aboard *Resolution*, and his fellow captain Tobias Furneaux, on *Adventure*, could plot both the latitude and longitude of any land they discovered in the Southern Ocean.

The ships sailed south from the Cape of Good Hope in the early southern spring. For the next three months, Cook dodged drifting ice and pods of whales, prowled Antarctic waters, and relentlessly searched for the mysterious *Terra Australis incognita*, even though Furneaux had been separated from him in a foggy ice field and prudently departed for a prearranged meeting in New Zealand.

In January 1774, his frail wooden ship dipped below 71° south latitude, the farthest south any humans had yet ventured. The floating ice had become a hazard and instinct told him unbroken sheet ice had to lie beyond the fog-bound southern horizon. Now the former North Country deckhand was certain he had solved the mystery of the temperate southern continent: If it existed, it was forever buried by glaciers and surrounded by impenetrable pack ice.

Cook and his crew wintered in the tropics, where they explored and charted vast numbers of islands unknown in the West. His ceaseless questing had sapped Cook's

On a third and final expedition to explore the North Pacific, Captain Cook's ships visited Kealakekua Bay on the island of Hawaii. Relations with local Hawaiians deteriorated, and Cook was killed on February 14, 1799, in a fray over a stolen British longboat.

health; although still in his forties, he was almost crippled by rheumatism and the adverse affects of his era's quack medicine. Still he would not surrender: En route home to England via Cape Horn, he again ventured far south seeking the elusive continent, even though as he noted in his log, he was heartily "sick of these high latitudes." After the southern hemisphere sailing season worsened, he finally turned north.

In the summer of 1775, Cook returned to a hero's welcome. In ill health, having already given his country a lifetime's service, Cook could have remained comfortably ashore. But again he rose to answer the explorer's summons. The Crown and the Admiralty, satisfied that no fertile southern continent existed, now wanted Cook to help answer the second major vexing mystery of their time: whether there was an ice-free Northwest Passage connecting the Atlantic and Pacific above the continent of North America.

On Captain Cook's final expedition, Cook's crew encountered polar bears during their unsuccessful search for the ice-free Northwest Passage around the top of the North American continent.

Cook, now a Royal Navy post captain, accepted the Admiralty's invitation to lead a two-ship expedition to the far northern Pacific, traveling aboard *Resolution* via Cape Town and the Indian Ocean, accompanied by the second vessel, *Discovery,* under Captain Charles Clerke. In January 1778, he sighted the lush green volcanic cones of an island chain that he marked on his chart the Sandwich Islands (today's state of Hawaii), the Earl of Sandwich then being First Sea Lord and commander-in-chief of the Royal Navy.

Throughout the long days of that Arctic summer, *Resolution* and *Discovery* probed the fjords and glacier-bound inlets of the Alaskan peninsula, once again among the hated high latitudes and icy fogs Cook had come to dread on the opposite end of the planet. In spite of unremitting efforts, Cook's explorers found no sea channel leading east, even though *Resolution* had sailed past 70° north latitude, well into the Arctic Ocean.

With the onset of winter, Cook and Clerke returned south to the island of Hawaii and anchored in the sheltered water of Kealakekua Bay. To the local people, the strange beings on the huge winged canoes could only be gods returned in mortal form. Cook had no idea of the native Hawaiians distorted perception of the British sailors.

However, the Hawaiians realized the seamen too were mortal when a sailor died. This opened the way for minor pilfering of sailors' possession which eventually led to the theft of *Discovery*'s cutter. Cook took a squad of armed red-coated marines ashore, intending to exert the necessary authority to retrieve the cutter, which was vitally needed for the next season's coastal exploration. A series of events ensued, ending in Cook being stabbed and beaten to death along with four of his men.

Captain James Cook, perhaps the most dogged, far-ranging explorer in history died at the age of 50 leading his final expedition, which proved that a Northwest Passage free of ice year-round did not exist.

In early March 1977, I stood at the rail of the research ship *Knorr*, watching a sea lion swim lazily across Discovery Bay on the Galápagos Island of Santa Cruz. We were anchored off the Charles Darwin biological station on the island's southern coast. The morning sky was cloudless, and the sun pleasant, but still I felt a slight chill to the air as if we were moored near one of the dry, treeless islands off the southern California coast, not virtually astride the equator, 570 miles west of South America. Here the cold, upwelling Humboldt Current sweeping north from Antarctica dominated the climate, overpowering equatorial humidity, so that the flat coasts of the volcanic islands were tangles of cactus, mesquite, and thorny brush. Only on the high steeper slopes of craters where the southeast trade winds collided with land did clouds form and rain fall reliably.

The Galápagos Islands' isolation and unique microclimate had triggered unusual and sometimes unparalleled evolutionary experiments. Here there were iguanas, land reptiles elsewhere, completely adapted for the marine environment, which subsisted on seaweed. The islands' six species of giant tortoises—*galapago* in Spanish—were some of the longest living creatures on the planet. Then there were the birds. Among the archipelago's 16 islands, there were unique flightless cormorants, flamingos, a bewildering variety of finch subspecies, and penguins.

But the spectacle of seeing little gray-and-white penguins swimming through these chill waters so close to the equator was only one of the bizarre sights we'd encountered here and on the surrounding sea bottom of the Galápagos Rift.

Our multidiscipline, multi-institution expedition had come to study the effect of plate tectonics on the fast-spreading seafloor of the region, using the towed camera sled *ANGUS* and manned submersible *Alvin*. The expedition's goal had been to investigate the undersea geology of the Galápagos Rift, a branch of the Mid-Ocean Ridge, in which we hoped to find hot springs. Not only did we discover them, but we also found ourselves confronting possibly the strangest marine biology that scientists had ever discovered.

During our first reconnaissance run down the Rift, a temperature sensor on *ANGUS* had revealed the presence of hydrothermal vents, fountains of superheated seawater spouting up from magma chambers below the lava seafloor. Photographs showed weird collections of unknown animal species surrounding these vents: huge white clams, giant tube worms with blood-red maws, and colonies of a completely unidentifiable creature that looked like an immense carnivorous dandelion out of some 1950s science fiction movie. Most startling, the nearby fresh lava was almost free of sediment, which meant there wasn't enough conventional nutrient to support such large and diverse pockets of life.

The newly discovered creatures existed on a biological foundation separate from the solar-photosynthesis circuit previously believed necessary to support life. Sampling the water from the hydrothermal vents, we discovered it to be rich in dissolved hydrogen sulfide, which apparently supported anoxic bacteria, the bottom of this unique food chain. In effect, we had stumbled on a new branch of evolution.

The Galápagos, of course, are closely associated with the founder of evolutionary science, the 19th century British naturalist and sometime student of geology, Charles Darwin. It was on these unique biological outposts that Darwin began to formalize his revolutionary theory that shook the underpinnings of conventional scientific thought and religious dogma 140 years ago.

One of humanity's great intellectual explorers, Britain's Charles Darwin, probably would not have developed his theory of evolution had he not been constantly seasick during the five-year voyage of H.M.S. *Beagle* and sought refuge ashore.

Following pages: **The Galápagos Islands were Darwin's laboratory of evolution. He discovered intriguing differences between the archipelago's flora and fauna as he visited the individual islands.**

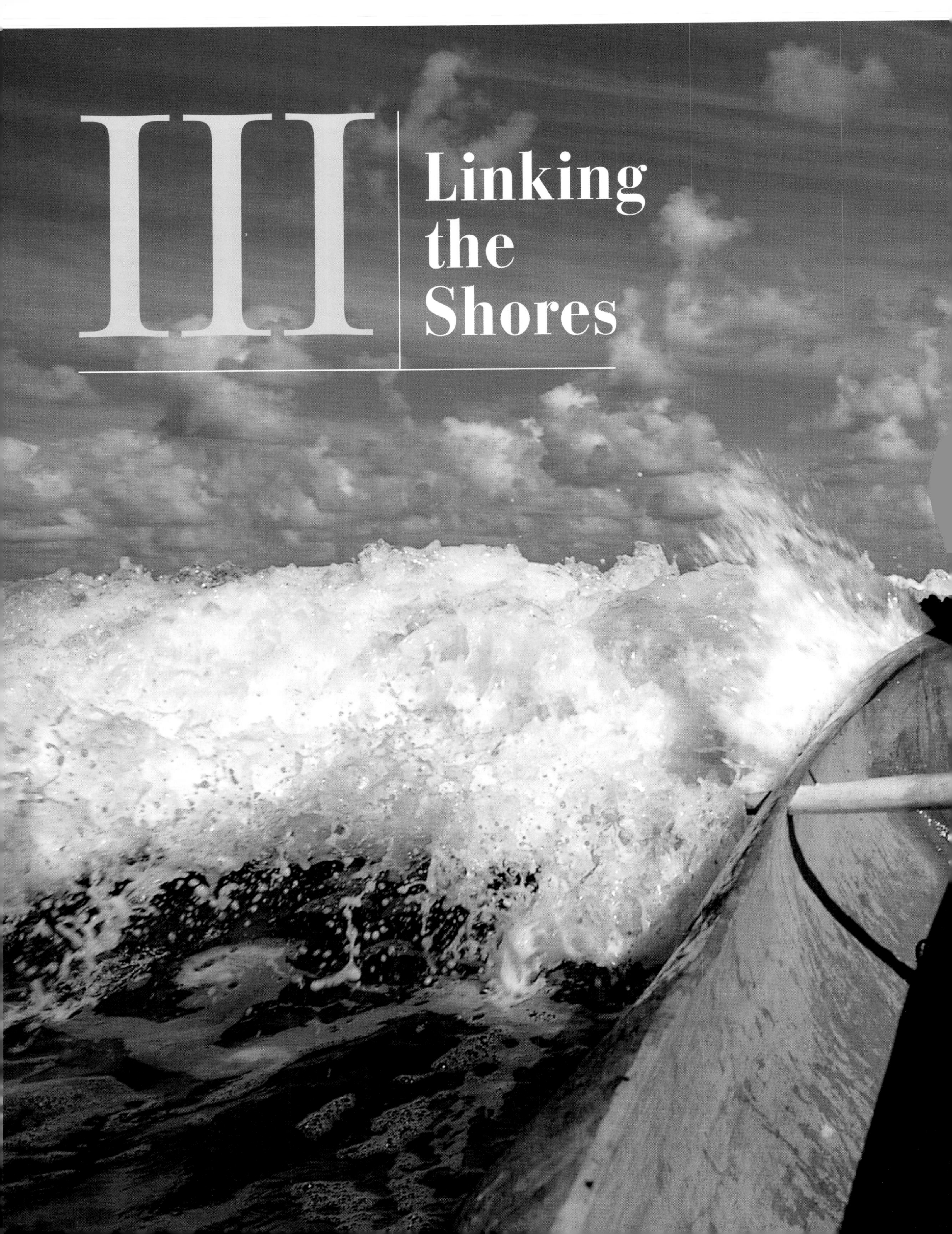

III Linking the Shores

Preceding pages: **An Andaman islander rides the prow of a dugout canoe through coral reefs separating mainland Asia and Indonesia in the Indian Ocean. It was aboard such vessels that questing voyagers long ago extended humanity's reach to Australia and beyond into the empty Pacific.**

Unlike any other nuclear submarine in the world, the *NR-1* has viewports. Although designed for classified military assignments, the *NR-1* proved to be an ideal research submarine for undersea archaeology, including our search for Roman shipwrecks on the Mediterranean seafloor between Tunisia and Sicily during our 1997 expedition.

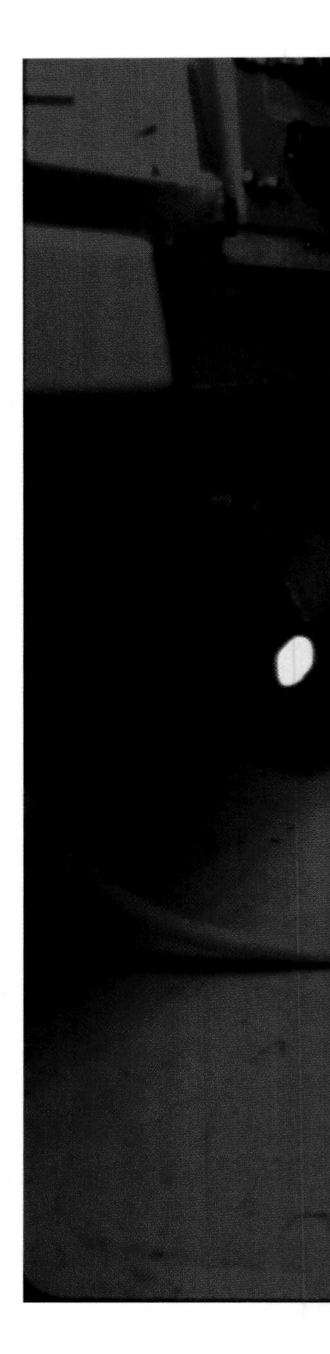

I lay comfortably in the narrow observer's station of the small U.S. Navy nuclear submarine, *NR-1*, my elbows resting on a rubber pad, my face near the armored-glass portal. The sub's piercing thallium iodide floodlights cast a spectral green radiance across the flat muddy bottom of Skerki Bank. It was early summer, 1997, two years before the Ashkelon expedition. We were operating, north of Tunisia and east of the Sicilian port of Marsala, below 2,000 feet—a depth that would have crushed even today's most advanced nuclear submarines.

But the *NR-1* was not your typical sub. With a tempered steel pressure hull and a nuclear reactor about the size of a trash can, the vessel could dive to 3,000 feet and stay down for weeks, probing the bottom with an array of high-tech sensors that included some of the most powerful sonars in the world, which could detect small distant objects. And once the crew had located the particular target they were seeking—often military hardware during highly classified missions—the *NR-1*'s manipulator arm could seize it in strong robotic jaws for storage in an external bin.

"Coming up on Site D," Lieutenant Commander Scott Swehla announced over the intercom from his seat in the control station just above me. He was maneuvering the sub with a computerized joystick, guided by images on video monitors and data from the sonar dome in the sub's hydrodynamic nose. Scott's control station reminded me of the pilot's seat in a space shuttle orbiter. Indeed, the neatly arranged crew quarters with tiny galley and pancake stack of bunks brought to mind a cross between a spacecraft and a high-tech RV.

As I watched, the monotonous gray mud of the bottom was suddenly broken by geometric shapes. Scott slowed the *NR-1*, and we barely glided ahead. Now I recognized the distinctive jumble of long-necked narrow and plumper terra-cotta amphorae, arrayed in a slightly oval pattern of about 50 feet that marked an ancient shipwreck.

"Bingo, Scott!" I called up to the control station. "We've got another one. Let's log the coordinates."

I had previously come to Skerki Bank in 1988 and 1989, on multigoal expeditions using camera sleds and ROVs to explore submerged volcanic seamounts with their associated hot-spring vents, and also to search for possible ancient shipwrecks in the restricted channel between Sicily and Tunisia. The Strait of Sicily was a logical point for this archaeological exploration: The narrow body of water connected the basins of the eastern and western Mediterranean, and also linked the shores of Roman Italy and the Imperial provinces in present-day North Africa.

The 1989 expedition, which was a shakedown cruise for our new ROVs, *Hugo*, *Jason*, and *Medea*, was an unprecedented success. An important and unusual aspect of this trip was our first live JASON Project satellite television links to twelve museums and research centers in North America. Hundreds of thousands of students were able to share in the frustration and jubilation of discovery as our research vessel, *Star Hercules*, rolled in the Mediterranean swell above the site where we had photographed amphorae strewn on the bottom the year before.

We deployed the sophisticated *Medea-Jason* system to conduct a thorough exploration. *Medea* was a kind of deep-sea eye, equipped with lights and video cameras that could watch over *Jason*, the collection ROV, to which it was attached by a neutrally buoyant tether. This technique protected *Jason* from the inevitable jerks and tugs on the long cable connecting *Medea* to the surface ship.

While the students watched via satellite TV, ninth grader Louise Jones, one of

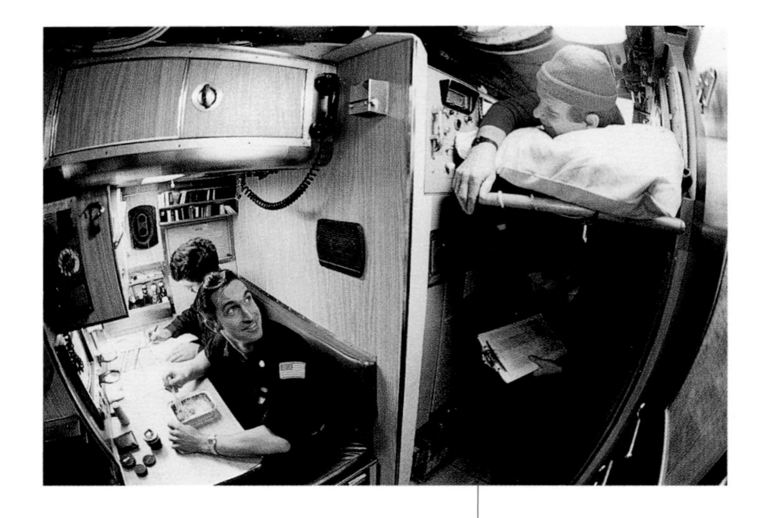

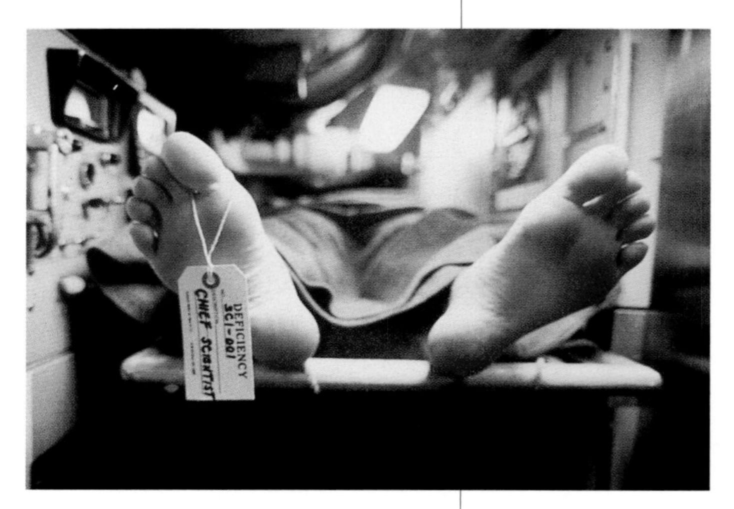

I peer down from my improvised pipe bunk on the *NR-1* and joke with crewman Buckley Bailey eating in the cramped galley. Despite tight quarters, crew morale remains constantly high.

Practical jokes break tension on the *NR-1*. Here the crew tag the toe of the chief scientist (me) with a "deficiency" notice during an earlier expedition.

their JASON Project counterparts, guided the two ROVs to the distant bottom. Soon *Jason*'s low-light video camera picked up a clump of amphorae of various shapes. Many were in mint condition; others were broken, hollow shells. As *Jason* skimmed across the field of amphorae, Dr. Anna McCann, an archaeologist specializing in Roman trade and artifacts at Trinity College, Dublin, leaned toward the video monitors in the control room. From the shape of the tan terra-cotta cylinders, she could tentatively date the shipwreck to the third or fourth century A.D., the height of commercial expansion during the Roman Empire. Professor McCann spoke to the students in North America, sparking their enthusiasm for the thrill of exploration. "Just imagine. You're seeing what no one has seen for almost 2,000 years."

Our preliminary video survey of the site suggested the faint teardrop shape of an ancient shipwreck. The flickering monitors showed the outline of a possible Roman merchantman's hull, just under 100 feet long. After we had carefully video-mapped and photographed the site, *Jason*'s pilot, Martin Bowen, hovered the ROV over the wreck, which we had decided to call the *Isis*, in honor of the goddess worshiped by seafarers across the pagan Mediterranean.

Shipbuilding was an art that slowly evolved in the Mediterranean from the coastal vessels of the Bronze Age, through the Iron Age Phoenicians' round boats such as *Tanit* and *Elissa*, to Admiral Hanno's fleet, and centuries later into the far-flung seaborne commerce of Pax Romana. Basically, however, ancient shipwrights followed the same "shell first" technique by which they started with curved rows of planking, their edges meticulously smoothed by an adz, with each pair of planks joined by locking mortises and tenons. The square, tablike tenons were pounded into the chiseled mortise holes and fastened by pegs at each end. Only when the entire plank shell of the hull was finished did the builder insert the ribs—a construction sequence for wooden vessels opposite that which arose in Medieval times and persisted into this century.

The classical mortise-and-tenon building technique produced extremely seaworthy ships, so tightly planked they needed no caulking. But relentlessly chipping away excess wood so that each plank edge precisely abutted its partner was also extremely wasteful. By the first century B.C., Greco-Roman shipbuilding had reached unusually sophisticated levels. Portions of a wreck lifted from the preservative inshore bottom sediments near the French Mediterranean port of Toulon revealed a double-plank construction, one hull shell inside the other. The outer skin of planks was sheathed with beaten lead sheeting onto which tarred flaxen fabric was tacked. A vessel so hulled could have carried heavy loads, perhaps thousands of amphorae filled with valuable wine or olive oil, and the entrepreneur merchant captain typical of the period would have sailed knowing that wood-boring worms would not dangerously weaken his ship. But the investment in labor and planks would have been enormous.

Indeed, historians believe that the rampant building of warships and merchantmen during the centuries of Greek and Roman dominion of the Mediterranean denuded many areas of their forests, particularly in the Aegean. The environmental expense of this technique can be traced in changing planking patterns: Classical Greek and Roman ships used large tenons, closely set; by the collapse of the Roman Empire the tenons were small and widely spaced, probably necessitating caulking.

Carefully employing *Jason*'s mechanical arm, we retrieved a small piece of wood from beneath the protective sediment line at the *Isis* wreck site. The wood proved to be white oak, which grows in Europe but not North Africa or the Middle East, so we concluded the ship might have been built in Roman Italy. After recovering a representative sampling of additional artifacts, I decided to not disturb the wreck site further on this expedition. The *Isis* had lain unmolested for 17 centuries; we had the site's exact coordinates, and we could always return to conduct a thorough archaeological "dig" once our robotic underwater excavation techniques improved.

The *Isis* artifact that most intrigued me was a small flattish terra-cotta oil lamp that *Jason*'s Cow Catcher delicately retrieved from the swirling sediment on our last day at the site. When we hoisted the elevator retrieval system on board, that little ocher oval sat dripping in the red netting between the pipe frames. Anna McCann used a piece of clean cloth to lift the delicate clay lamp and cup it in her open palms as she explained to our distant JASON Project students that the distinctive rounded heart shape dated the artifact from the fourth century after Christ.

Aside from a single calcified worm shell near the stubby handle, the lamp was free of marine life. But the terra-cotta surface was stained black from the flame of the olive-oil wick that had burned so long ago. I pictured a young sailor, perhaps the son of the captain, assigned to cook, crouched in the low galley in the flickering light of that lamp, preparing the crew's final meal of bread and fish stew. Then catastrophe—probably a savage storm that raised dangerous hollow seas like those that had sunk the Phoenician vessels—overcame the ship. The hull was swamped, the tiny flame extinguished, and the *Isis* sank to the bottom carrying that delicate clay lamp.

Eight years later, during our search for ancient wrecks aboard *NR-1* west of Sicily in 1997, Lt. Com. Charles Richard, the sub's skipper, and his crew were locating a promising wreck site almost every day. After Richard had carefully logged these new coordinates, I took him aside in the sub's narrow crew quarters. "Charlie," I said, pointing to the small rectangles neatly penciled on our survey chart, "keep up the good work, but I'm going to head topside to the *Carolyn* and switch to the ROVs."

That afternoon, I transferred to the well-equipped research vessel *Carolyn Chouest*, which I had often used on previous expeditions. My team from the Institute for Exploration and the Woods Hole Oceanographic Institution included Dana Yoerger, who understood the intricate workings of the bright yellow, 1.5-ton ROV *Jason* better than most Kentucky Derby trainers knew their thoroughbreds. Once more, Anna McCann headed up the archaeology and conservation team.

Using the *NR-1*'s sonar data, we zigzagged back and forth on the basic north-south course between Carthage and the port of Ostia, near Rome at the mouth of the Tiber River. This survey area included the convergence of sea routes from not only the North African bread basket of Imperial Rome, but also the logical courses of ships transiting between the eastern and western Mediterranean. In all, we ultimately located eight wreck sites, but decided to survey, map, and sample for

The Roman shipwrecks we discovered in the Mediterranean carried the empire's wealth in the form of wine, olive oil, and fish sauce stored in terra-cotta amphorae. Archaeologists can date each shipwreck by the characteristic shape of its amphorae.

***Following pages*: The nuclear-powered *NR-1*, capable of diving to 3,000 feet, employs powerful sonar, a specialized lighting system, and a robust manipulator arm to perform its unique missions. It has been rightfully called an undersea space shuttle.**

NR-1
U.S.NAVY

artifacts only the oldest five, which we believed dated from the Roman era, spanning a period from the first century B.C. to the fourth century A.D.

The generally narrow corridor in which these wrecks were distributed confirmed our hypothesis that there was a thriving ancient trade route between Carthage and Ostia, and also that vessels heavily laden with trade goods from the eastern Mediterranean rounded the seaward side of Sicily—avoiding the often storm-bound Strait of Messina—on their route north to Roman ports. After eight weeks of painstaking video-mapping and still-photo mosaic work, as well as recovering key selected artifacts to date the wrecks and trade routes, we verified that these Roman ships carried goods from both the eastern and western Mediterranean, as well as an abundance of merchandise from the European and North African shores of the Roman Empire.

Working in the *Carolyn Chouest*'s laboratory one evening, Anna McCann laid out a black-glazed ceramic plate from central Italy dating from the first century B.C., and pointed to two squat, cream-colored amphorae with unusual flat bottoms, which lay in a vat of preservative fluid to prevent too-rapid drying. "The plate is Campanian," she said, "mid-first century B.C. But those amphorae are distinctive. They're from Gaul, and were used to transport either wine or beer. They just might be the earliest evidence we've ever found of wine export from Gaul." The clay of the ancient wine containers was so fine that they did not require the internal pine-pitch coating needed to seal amphorae. We'd found possible evidence of an early divergence between the rough, tarry retsina wines of the eastern Mediterranean and the more refined vintages of southern France.

Arab shipwrights in Dubai on the Persian Gulf patiently build a dhow using techniques unchanged for centuries. Their tools include the adze and the bow drill for boring holes in planks.

An Arab dhow plies the aquamarine waters of the Persian Gulf. Although diesel-powered craft are now numerous, the lateen or Latin, rig still prevails.

The artifacts and amphorae were traceable to the factory-scale pottery run by the Sestius Family and had convinced Anna that the ship was undoubtedly loaded at the Roman port of Cosa on the Tyrrhenian coast of Italy, before heading south into the open Mediterranean.

The heaviest laden of the merchant ships, dating from the first century A.D., carried an amazingly diverse cargo, including rough-cut granite blocks and columns, already shaped as if for some Imperial contractor's building site. The ship also carried large amphorae, which had probably held wine, olive oil, and the pungent fermented fish sauce, garum, so prized by the Roman gourmet. The cargo hold contained hundreds of sets of clay pots and plates as well, apparently all packed for delivery at a long-forgotten pottery. The heavy anchor had a valuable iron arm and flukes rather than the cheaper lead anchors of the period, indicating this vessel was owned by a prosperous merchant.

But wealth, a stout hull, and well-woven sails could not protect these Roman sailors from the cruel reality of their risky profession. The Mediterranean, with its shoals and lurking rocks, could swallow ships even during good weather. *Jason* and the *NR-1* had located elongated "amphora alleys" scattered across the seafloor in distinctive patterns indicating terrified mariners attempting to lighten their loads by dumping precious cargo overboard, one heavy terra-cotta container at a time. The Romans carefully delineated the optimum sailing season as being from the end of May to mid-September. In dire necessity, merchants might chance sailing from the second week of March to the second week of November. But only fools, or captains motivated by greed or Imperial decree, left port in winter. However, there were no doubt enterprising captains who dared to skirt the edges of the season, delivering military dispatches or especially valuable cargo such as Indian spices transshipped from Alexandria. Hundreds of these bold seafarers must have been lost.

The onshore winter monsoon whipped dust across the road as my old friend and University of California at Santa Barbara Geochemistry professor, Clifford A. Hopson wheeled the Land Cruiser off the rutted track and stopped beneath the sparse shade of a stunted acacia. It was January 1981, and we had come to the Sultanate of Oman on the southeastern corner of the Arabian Peninsula to study the floor of a long-extinct ocean. Instead of searching with deep-tow camera sleds or diving in submersibles, however, Cliff and I were conducting a traditional geologist's field trip, with hiking boots and sample hammers.

During a 1980-81 re-creation of the voyage from Oman to China of the legendary Sindbad, Indian fishermen at Bayport on the Calico Coast sew a new suit of lateen sails for the expedition's replica medieval dhow *Soar*.

Cliff was one of the world's leading experts in ophiolites, huge sections of ancient ocean bottom that tectonic forces had heaved up intact. He pointed out a dominating ridge to the west. "That's Al Jabal al Akhdar," he said, gazing toward the darkly brooding mound that stood above the treeless volcanic ridge of the Al Hajar mountain range, "'the Green Mountain.'" Even from this distance, I noted the presence of copper-bearing minerals. The mountains and blocks of pillow lava heaped before us had been seafloor 95 million years ago, part of the mid-ocean ridge in the Tethys Sea. Then the Arabian and Eurasian Plates had collided and that seafloor had slowly been elevated to become dry land. Before us was a well-preserved fossil record of that ocean bottom's multiple layers.

By the end of the afternoon, I was gritty with lava dust and my shoulders ached from prying boulders loose, but I was pleased with the effort. As much as I enjoyed working in a submersible like *Alvin*, it was still fun to hike among mountains and get my hands dirty.

Later, Cliff drove south along the coastal road, back toward the capital, Muscat. Now the wind was churning the turquoise Gulf of Oman. Far offshore, the dark shapes of supertankers moved silently through the distant swell, seemingly untouched by the force of the strengthening northeastern monsoon. But nearer the beach, I saw traditional Arab dhows cresting the mounting swells as they sailed downwind for shelter within one of the Sultanate's modern concrete breakwater ports. Their triangular lateen sails, gleaming with spray, had a timeless quality.

A sign pointed to a turnoff and I caught a glimpse of derricks and masts, probably the fishing port where those dhows were headed.

"That's Suhar," Cliff said, "reputed birthplace of Sindbad the Sailor."

As I watched the wind work its ever changing patterns on the water, I thought of Sindbad, a hero from *The Thousand and One Nights*, who was every bit as courageous and resourceful as his Western counterpart, Ulysses. The mythical Sinbad had sailed from the Persian Gulf in the eighth century A.D., making seven hazardous voyages to Asia. After many adventures, he had returned with his fortune, and also, like so many legendary explorers, bearing the priceless cargo of hard-won knowledge.

The Gulf's real seafaring merchants' mirrored this great achievements. They carried back the ancient secret of the monsoons, knowledge that had been well known to the Romans, then lost when their Empire collapsed.

Retracing Sindbad's voyage, modern sailors handle torn sails during a brisk squall in the South China Sea.

Studying maritime history, I've learned that many prosperous Roman merchant sailors during the tranquil decades of the third century A.D. did not have to contend with the rocky hazards and unpredictable weather of the Mediterranean.

Instead they ventured forth on a generally less hostile body of water—the Arabian Sea—relying on the reassuring annual clockwork of the monsoons to carry their square-sailed vessels before the wind from Imperial outposts in Egypt to exotic spice ports of India. With the Empire nearing its pinnacle, several million affluent Roman citizens living in cosmopolitan cities stretching from Britain to the Middle East demanded luxury goods, notably spices such as black pepper and cloves, to enhance their increasingly sophisticated cuisine.

This demand spurred the enterprising seaborne Roman traders to attempt the 3,000-mile round-trip voyage from the Red Sea to spice ports on India's Malabar coast. The open waters of the Arabian Sea must have initially presented a daunting challenge to the first Roman sailors (who in fact were probably Alexandrine Greeks or merchants from the Levant) attempting the passage. But countless generations of local fishermen and coastal merchants would have attested to the regularity of the southwestern (summer) and the northeastern (winter) monsoons. And scholars now believe adventurous Chinese traders, who had first reached India and the Persian Gulf from ports at the mouth of the distant Yangtze River, carried the secret of the monsoon's true ocean-spanning reach to the Red Sea frontier of the Roman Empire.

These winds, named for the Arabic *mawsim*, season, were related to the prevailing summer northwesterlies of the Mediterranean that fed into an annual low-pressure system stalled over Persia and the baking Subcontinent by the high wall of the Himalaya. The rain-bearing southwestern monsoon blew steadily across the Arabian Sea and upper Indian Ocean, one hot day of low gray scudding cloud following another for months, always with the moderate humid ocean wind flowing from the Horn of Africa to the Malabar coast. In the winter, the pattern was reversed. Cold high pressure spilling from frigid central Asia over the Himalaya was drawn into a wide low-pressure system over equatorial Africa. This was the gusty wind I witnessed on the coast of Oman.

The Roman sailors who first dared this voyage must have trusted the strength of their well-planked ships and the judgment of their captains. I can imagine what they experienced, sailing with the cloying moist wind behind them, neither the sun nor stars visible through the thickening overcast. The helmsman at the long steering oar on the curved side of the stern bulwark would have had no fixed point of reference to guide him. The warm tropical seas beneath the bow would have flashed with the eerie green fire of bioluminescence much more intense than that seen in the Mediterranean. The only constants were the steady wind and the rolling swell. Sailing at an average speed of five knots, the voyage would have taken at least two weeks beyond the sight of land, longer if prudent captains reduced sail at night.

As these explorer-merchants became bolder, they extended their voyages around to the eastern shore of India and even into the South China Sea. There are third century A.D. accounts of Roman ships visiting "Kattigara," which was probably Hanoi in today's Vietnam. Archaeologists have unearthed Roman coins of this period in Indochina, and Roman pottery was in common use on the coasts of India. At the return from these voyages, deeply laden with invaluable cargoes of pepper, cardamom, cloves, golden white elephant ivory, and heavy ebony logs, the audacious merchant sailors would have reaped the rewards of their daring. This monsoon-borne commerce, conducted on well-found vessels similar to those my expeditions explored on the Skerki Bank, continued until the collapse of the Roman Empire.

Within a few centuries, the dhows of the Muslim sailors who now controlled the Persian Gulf and the Red Sea had taken over the monsoon spice trade. The legendary Sindbad and his real-life counterparts were the masters of the Arabian Sea. Their vessels generally had narrower hulls than Roman merchantmen and relied on a large triangular lateen (Latin) sail of eastern Mediterranean origin, ideal for maximum speed downwind in the monsoon. Outbound from the Arabian Peninsula, the Muslim sailors evangelized zealously for their new religion, eventually carrying Islam all the way to the distant Indonesian archipelago. There they were poised at the Pacific Ocean's vast expanse, which encompasses over half the planet.

The relentless southeast trade winds swept the grass of the hillsides in undulating waves. Today the island belongs to Chile and is home to a few hundred Spanish-speaking Polynesians who raise sheep and cater to the occasional tourist who finds their way here. Jean Franceteau and I hiked steadily toward the unearthly vertical shapes looming in the distance.

It was January 1983 and we were on Easter Island, one of humanity's most remote outposts. Jean and I had flown in from Chile, 2,300 miles to the west, to join the French-American expedition exploring the tectonic formations of the surrounding East Pacific Rise using the research vessel *Nadir* and the submersible *Cyana*.

Monumental statues of dark volcanic tuft, each capped with a massive pukao topknot, stand on *ahu* terraces on remote Easter Island. The obsession for building these statues eventually ignited disastrous war between the island's classes.

The ship was anchored off the island's main village, Hanga-Roa, on the sheltered west coast.

With a day free before sailing, several of us had come to see the famous monolithic stone statues. But we weren't prepared for the sheer size of these monumental artifacts carved from the island's dark volcanic tuft. They stood on quarried tuft-block *ahu* burial terraces dug into the hillsides. What looked at a distance to be slim cylinders perhaps ten or twelve feet high, proved to be immense stylized heads with blade-like noses, elongated earlobes, and back-tilted foreheads. Each head was precariously capped by a gigantic red tuft *pukao* topknot, raising the total height of these statues to over 30 feet.

"This one weighs almost 60 tons," Jean said, consulting his guidebook.

I gazed down the rank of statues. Behind us the grassy volcanic hills sloped away to the bluffs above the empty blue Pacific where seabirds wheeled. The statues' blind eyes revealed nothing of the violence this lonely island had witnessed.

But in the crater called Rano Raraku we could piece together the tragic climax of the Polynesian culture that once thrived here. Before us a gigantic unfinished statue, almost 70 feet long, half hacked from especially desirable grayish yellow tuft lay on its side, still surrounded by the stone picks, wedges, and hammers of the work force who had suddenly abandoned their task centuries before.

This mute evidence of a violent upheaval coincided with other unmistakable signs of strife. Hundreds of statues had been toppled from their ahus. Family wealth

Twin-hulled Polynesian pahi, guided by expert *fa'atere* sailing masters and capable of voyaging thousands of miles, spread human culture to the remote corners of the Pacific in humanity's most epic voyages.

size, frequency, and duration of the swell told them if islands lay ahead that were interacting with the prevailing winds.

In late afternoon or at dawn, the shallow, white-sand lagoon inside a distant reef could cast a gleaming shimmer on clouds. And, as I have often seen in the Pacific, the steep, jungle-clad ridges of seamount islands trapped moist winds to form distinctive cloud caps visible for more than a hundred miles in the right conditions. The larger islands often made their own weather when thermal updrafts created thunderheads. The far-off glare of lightning that seemed stationary, unlike swift ocean thunder squalls, could also indicate a mountainous shore over the heaving gray horizon. Fishing birds such as boobies roosted on islands, but flew long distances offshore to feed. Following them home was a wise expedient.

By these complex skills, acquired over centuries and passed on from one generation of fa'atere to the next through long apprenticeships, the Pacific voyagers were able to sail amazing distances. For example, the passage between Tahiti and Hawaii of over 2,000 miles—probably first successfully completed around 400 A.D.—was a feat of open-ocean seamanship without the benefit of guiding monsoons not equaled by European explorers until a thousand years later.

Were these brave Pacific sailors true explorers, or migrants driven by economic necessity or the mundane misfortunes of weather? Was population pressure and warfare a factor in driving refugees from their native islands? These are difficult questions to answer, and probably all these conditions prevailed at one time or another. But there is also an ancient Polynesian tradition called "*imi fenua*," which can be translated as an honorable, but insatiable "searching for lands." This certainly suggests that conscious exploration, not random seaborne wandering, or desperate flight for refuge drove many voyagers of the Pacific.

The questing journeys were usually uncomfortable, and always perilous. Unlike the deeper-hulled European ships, the Polynesian double canoe could not provide adequate protection from torrential rain and gale-driven spray. A pahi had no fo'c'sle, only a flimsy shelter of woven palm fronds. The sailors had capes of braided grasses or beaten bark cloth, but these primitive garments were hardly as waterproof as the

oilskins European seafarers had developed in the Roman era. Death from prolonged hypothermia was no doubt a specter that stalked the Polynesian sailors. The other major danger was the weather. I've seen savage tropical squalls in the Pacific, but fortunately from the snug superstructure of an oceangoing ship like *Knorr* or *Melville*. When these squalls struck at night with their howling wind and horizontal rain, I imagined what it would have been like to be out there in the lightning-slashed darkness on a Polynesian outrigger. The region is also haunted by typhoons, the name we give to hurricanes in the Pacific. These are the ultimate tempests, which no doubt devoured countless families of these seafaring pioneers. But the people of Oceania, propelled by the questing human spirit, carried mankind, on these precarious craft, to the last unoccupied land on Earth other than Antarctica. *continued on page 88*

Following pages: **Polynesian explorers reached the Marquesas Islands in the South Pacific after centuries of hazardous exploration. Clouds forming on mountaintops often guided these sailors to landfall, but the voyagers still had to find safe passes through the coral reefs surrounding the islands.**

LEARNING TO CROSS OPEN WATER without the aid of landmarks was one of mankind's greatest accomplishments.

However, navigation was not an easily acquired skill. The coastwise Pleistocene sailors of the Indonesian and Solomon islands were stymied by the yawning blue emptiness of the South Pacific. And it took centuries for Lapita voyagers to perfect the art of the earliest *fa'atere* sailing masters.

After the fall of the Roman Empire, ocean voyaging was limited to the Polynesians and, in the Arabian Sea and Indian Ocean, to the Muslim traders and occasional Chinese merchants riding the monsoons. These seasonal winds were so steady in their direction that sailors could navigate without charts or instruments. But the Arabs and Chinese of the first millennium A.D. did steer by stars when the monsoon sky was clear.

Direction: In the early 12th century A.D., Chinese and European mariners began to use the first major navigational instrument, the magnetic compass. Although it was long believed the compass was invented in China, then spread quickly to other seafaring regions, there is now strong evidence that the critical instrument was independently, and almost simultaneously, developed in China and northwestern Europe.

Both Chinese and medieval European accounts describe a crude device in which a needle that had been magnetized from a lodestone (unrefined iron oxide) was floated on a reed or wooden chip in a bowl of water. The needle swung to a north-south alignment. Europeans saw the compass needle as pointing north to the Pole Star (Polaris). By the early 14th century, more advanced compasses, in which the magnetized needle was fixed to a central pivot, were in wide use.

The mercantile Genoese and Venetians equipped their ships with even more elaborate compasses in which the eight directions of the prevailing Mediterranean winds were marked on "cards" attached to the instruments' bowls or frames. Ironically, the millennia-old term "orientation," referring to the east where the sun rose and Jerusalem was located, persisted even though mariners now used north as their principal point of reference.

Zheng He's vessels all carried intricately made compasses, as did the far-ranging ships of the Portuguese mariners who reached Asia 80 years after the voyages of the treasure fleet.

Latitude: Estimating latitude—the distance above or below or the equator—also required ingenuity. Polynesian fa'atere observed that the farther they sailed north or south, the higher or lower certain stars peaked at their nightly zenith. Medieval European sailors used an uncomplicated cross staff to simultaneously align Polaris at the tip of the instrument's moving arm and the edge of the horizon at the other. The cross staff's fixed member was inscribed with numbers representing the observer's latitude. This simple but effective instrument served for centuries.

Medieval sailors also adapted the Classical Greek astrolabe, consisting of a movable pointer on a disk, to obtain latitude. Greek astronomers had employed the astrolabe to mark the passage of stars and planets. Beginning in the Middle

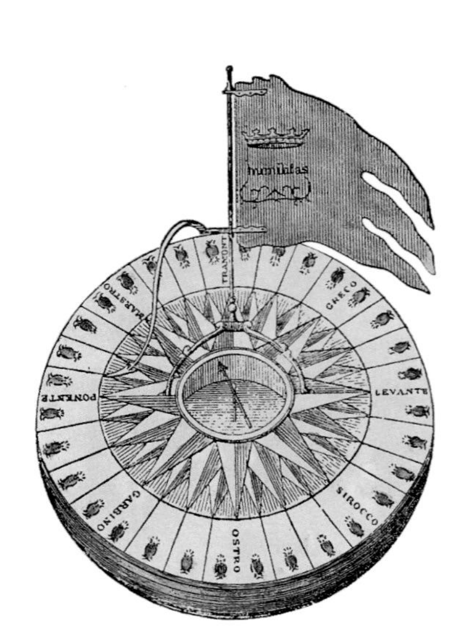

This Renaissance mariner's compass contains a magnetized needle which pivots freely. The surrounding circular card displays multiple directional points, which have been replaced by degrees in modern compasses. The card of this instrument also shows the prevailing Mediterranean winds such as the southeast sirocco.

Ages, instrument makers etched fine steel or silver astrolabes with precise calculations of the sun's declination, the seasonal progress of this celestial guidepost north or south of the equator. Equipped with an astrolabe, a mariner like Vasco da Gama could bravely sail the 20,000-mile voyage from Portugal to India and return.

By the 18th century, the cross staff and astrolabe had given way to the more elegant sextant. This instrument used mirrors and smoked glass filters and a small movable telescope attached to a curved limb. A navigator obtained the precise height of the sun, moon, star, or planet by "shooting" the celestial object—fixing its reflected image in the sextant's mirror—and swinging the object down to the ocean horizon. Reading the position of the telescope on the graduated edge of the limb provided the celestial body's altitude above the horizon. Two or more sextant shots could be joined by trigonometric calculation to provide exact latitude anywhere on the ocean.

As celestial navigation became more common, European seafarers needed better knowledge of the major stars' relative positions throughout the year. That ancient wisdom had been first obtained by astronomers in the Fertile Crescent, amalgamated by the Alexandrine Greeks, and conserved by the Arabs in such centers of learning as Damascus and Cairo. This is the reason that the principal navigational stars in modern nautical almanacs often have Arabic names. Betelgeuse, for example, the most prominent star in Orion's belt, comes to us from the Arabic *bat al-dshauza*, "the giant's shoulder." Aldebaran, a large reddish star, is called "the Fol-

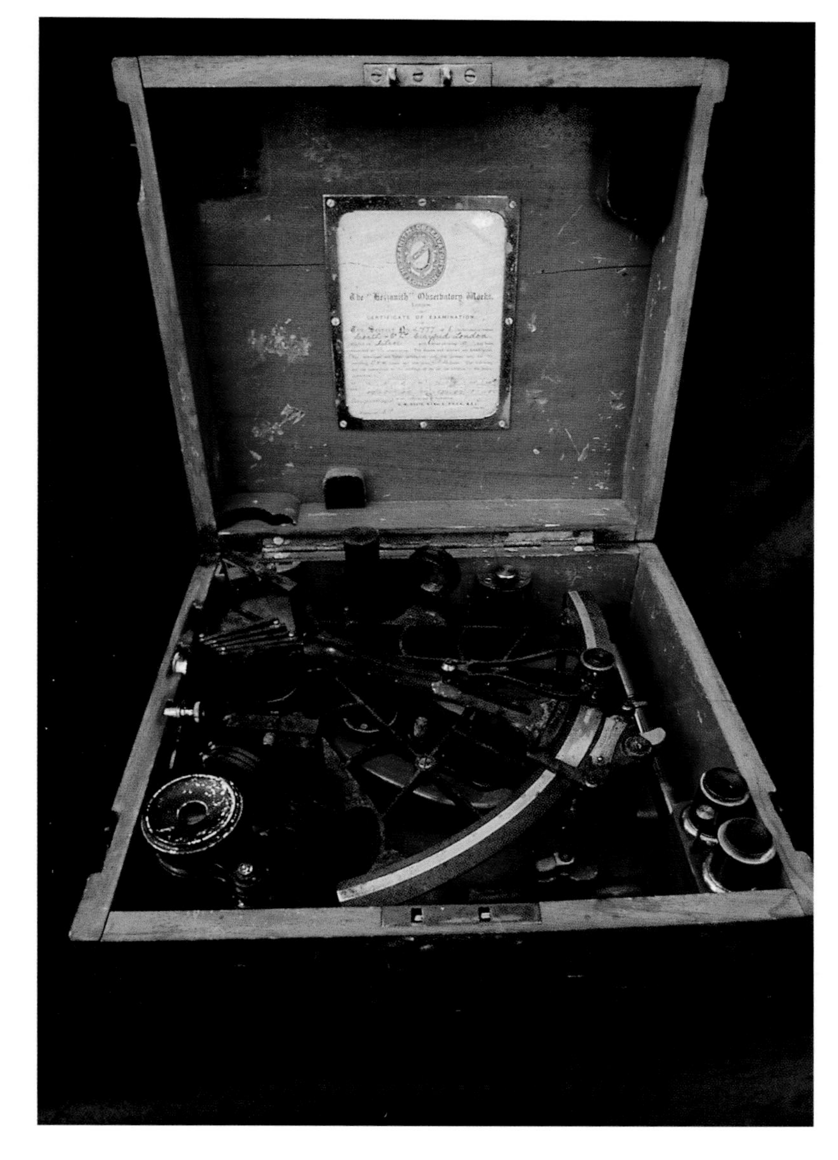

One of John Harrison's 18th century Chronometer No. 4 models (left), which successfully solved the "problem of longitude" that had long vexed mariners. A precision sextant (right) evolved from the astrolabe and cross staff. A sextant measures the height of a celestial body, allowing voyagers to plot their distance north or south of the equator.

lower" in Arabic. It rises from the ocean horizon following the dazzling star cluster of the Pleiades.

Longitude: How to judge longitude (the distance east or west of a north-south meridian baseline) stymied mariners for centuries. By the 1400s, the nations venturing forth onto the world's oceans had revived the Classical Greek understanding that Earth was a spherical planet, a globe, which could be divided into parallels of latitude and meridians of longitude to provide geographical coordinates. Geographers also recognized that Earth turned one complete revolution, 360 degrees, every 24 hours, or 15 degrees each hour. By comparing a ship's "local" time—usually noon, when the sun was at its zenith—with noon at a known meridian, the captain could judge how many degrees of longitude the ship had sailed east or west of that meridian.

But how to carry accurate time? For the British Admiralty, the "problem of longitude" became critical by the early 18th century. Ships sailing west to the Caribbean or east to India were foundering on rocks because their captains could not plot longitude. Parliament created a Board of Longitude in 1714 that offered a staggering 20,000-pound prize to the person who presented a practical means of solving this problem. Some mariners advocated complex astronomical methods, measuring angles between the moon and stars or tracking the progress of Jupiter's moons. But John Harrison, a Yorkshire clockmaker, was convinced the secret of reliably finding longitude lay in an accurate timepiece. Pendulum clocks, however, did not provide accurate time aboard a heaving, pitching sailing vessel.

Harrison labored for years building large spring-driven chronometers that did not depend on a pendulum. Only in 1762 did his small Chronometer No.4 meet the Board's strict requirements: A ship carrying the small instrument to Jamaica plotted its longitude with an error of less than five seconds (a fraction of a degree).

Because of the British government's sponsorship of the project, the Royal Observatory at Greenwich, on the Thames near London, became the site of zero longitude, the prime meridian, at least for the British; the Spanish used Cadiz.

Speed: Estimating speed sailed is also important for seafarers, especially those approaching a rock-bound coast or coral-reefed island downwind. By the late Renaissance, European mariners were regularly "casting the log," a wooden wedge, its tapered bottom weighted with lead. Attached to a long line marked every 47 feet with knotted cloth, the log was dropped over the vessel's lee side (away from the wind). The weighted log remained stationary as the ship moved ahead. By counting the number of knots that reeled out while a 28-second sandglass emptied, the ship's master could judge his speed. The term "knot" referred to one nautical mile (6,080 feet) per hour, which equals one minute (a sixtieth) of one degree of latitude.

Charts: Nautical charts were developed independently in the Mediterranean and China. The Classical Greeks drew charts showing parallels of latitude anchored on major trading ports such as Rhodes or Alexandria and extending east through the Mediterranean. Chinese chart makers mapped Asian waters near the continent, and later extended their reach with Zheng He's voyages. In the Mediterranean, the charts of medieval and Renaissance cartographers had complex webs of "portolans," compass-course lines that showed the best course to steer from one port to another using the season's prevailing winds.

As seafarers developed the ability to precisely plot their position and speed, they drew more accurate charts. Those describing phenomena such as the trade winds and helpful

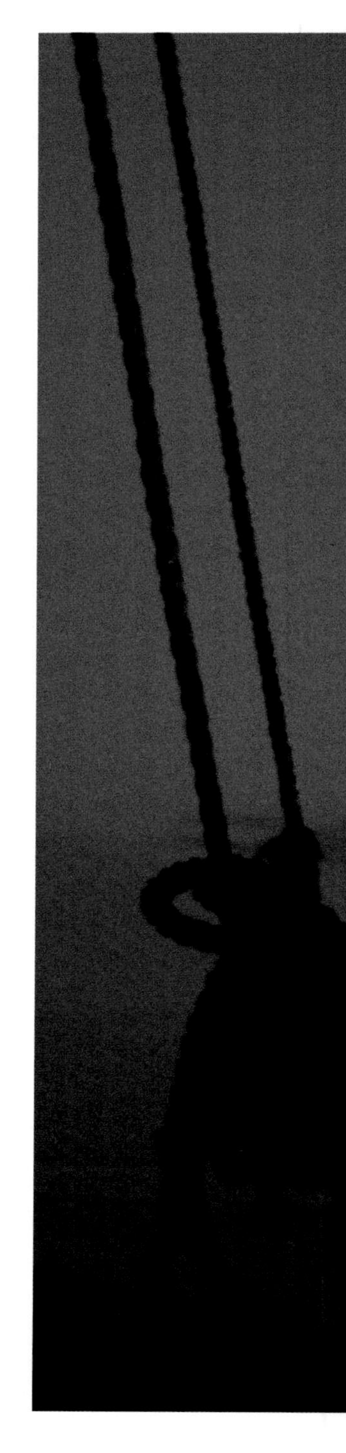

During the recreation of the legendary Sindbad voyage, explorer Tim Severin (left) uses a *kamala*, a type of sextant, to take the vessel's latitude. Aligning the wooden rectangle with the horizon's rim, he grips a string, knotted at lengths indicating the latitude of known ports, with his teeth. Although primitive, the instrument proved effective.

currents became virtual state secrets: Portugal, for example, guarded its knowledge of the trade winds for decades. Charts that plotted hidden hazards such as submerged rocks or revealed vital local knowledge such as a narrow passage through a reef were priceless. Among his many accomplishments, Captain James Cook is perhaps best remembered by mariners of later generations for his meticulous surveys and chart making. In fact, the finely engraved British Admiralty charts that evolved from Cook's pioneering effort are still widely used. Many a far-ranging yacht sailor has had the pleasure of unrolling an Admiralty chart of a South Pacific island and reading that the soundings of the lagoon and the elevations of the green ridges above were first taken by the officers and men of H.M.S. *Endeavour.*

I stood at the window of my hotel, savoring the air-conditioning as I watched the sunset on Hong Kong harbor. It was August 1992, and I had just spent several weeks in the Solomon Islands, investigating 13 lost Japanese and Allied warships of the pivotal battle of Guadalcanal. Working with the U.S. Navy, our expedition had used the ROV *Scorpio* (a cousin of *Jason*) and the submersible *Sea Cliff* (a sister vessel of *Alvin*) to explore these sunken wrecks in grimly named Iron Bottom Sound.

I was en route to Beijing, invited to discuss marine archaeology with the Chinese government. They wanted help in finding more of the ancient stone artifacts that fishermen had recovered from the Yellow Sea between China and Korea. I was interested in mounting an expedition to the waters off nearby Macao, where a fleet of imperial junks had sunk in 1276. Maybe we could do some horse-trading in Beijing.

Meanwhile, I was engrossed with Hong Kong after the near wilderness of the Solomon Islands in the South Pacific. The Victoria Central waterfront was a wall of glass high-rises mirroring the grapey afterglow of the harbor, which in turn reflected the sun setting behind Lantau Island. A bewildering collection of ships and boats crisscrossed the water between Victoria and Kowloon. There were little sampans, each skulled by a single fisherman or *wallah-wallah* peddler, fading into the dusk, wide ferries crammed with passengers, rusty tramp steamers, and hulking Japanese container ships that looked too top heavy to be stable. Across the harbor, the dark gray mass of a big warship, probably a U.S. Seventh Fleet cruiser, rode at anchor.

But the vessels that most intrigued me were the junks. I saw little fishing-family junks, sailing close-hauled against the humid southern monsoon, laundry flapping on the rigging. There were big, dark-hulled cargo junks with high sterns and powerful diesel engines, plowing confidently through the chaos of converging traffic. And on the edges of shipping, were the near derelict junks of uncertain age, their teak-planked hulls patched and splayed, yet still serviceable, reduced to carrying mounds of construction rubble or garbage. As the last sunset colors dissolved and moonlight competed with neon glare on the waterfront, I glimpsed the silhouette of a big seagoing junk against a distant channel. The vessel hoisted sail and heeled slightly into the steady wind, growing smaller by the minute. The sight of this well-found junk, her sails trimmed tautly in the monsoon, cresting the swell as she sailed into the darkness beyond the safety of the harbor was incredibly evocative of China's ancient maritime tradition.

Many maritime historians believe it was Chinese sea traders who brought the secret of the monsoons to the Roman world. For several hundred years, Chinese merchant vessels cruised down the Malay Peninsula, through the Strait of Malacca, and on to India. These ships, which we now collectively call junks (derived from the Portuguese *junco*), were superior to the Roman merchantmen and the Arab dhows that followed them. The classic seagoing Chinese junk often had a triple-layered hull of precisely edged planks, secured, with no caulking, by hardwood dowels. Unlike the clumsy, side-mounted steering oar of Mediterranean vessels, junks used a large central rudder fixed on a movable sternpost. A junk carried multiple masts, which permitted a great variety of sail combinations to suit almost any weather. The sails were woven rattan, stiffened by long, resilient bamboo battens running hor-

A Chinese junk plies Hong Kong harbor. These hardy vessels have sailed Asian waters for millennia. The huge junks of Admiral Zheng He's Treasure Fleet carried out the most far-reaching explorations of the early 15th century. Had China continued exploring, it would have become the world's dominant seafaring nation.

izontally across the entire width, and could be trimmed to take the wind ahead or behind. Junks had deep hulls divided into strong compartments, allowing them to carry impressive cargoes. The ships were flat-bottomed and could be beached to load or discharge their goods. These fast, well-found ships had no equal for centuries.

China, however, remained a primarily continental power until the early 15th century, even though its seafaring merchants had carried home tales of rich lands that could be plundered if the emperor only decided to exert naval dominance. The Ming Dynasty's court and government were in Nanking, a port on the Yangtze estuary. But the throne faced the vast inland plains and mountains. It was not until 1402 that a new emperor named Zhi Di called on his most trusted lieutenant, the Muslim eunuch Zheng He, to pursue Chu Yunwen, the deposed emperor, who was thought to have fled overseas.

I've always found Zheng He one of the ablest and most energetic seafarers in history. He meticulously prepared before setting forth on his first voyage. As High Admiral, he assembled over three hundred large armed junks and smaller ships carrying almost 30,000 crewmen and soldiers, an armada on the scale of Carthaginian Admiral Hanno's fleet. They sailed in mid-1405, calling in Indochina, then probing coves and anchorages on the Malay Peninsula in search of Chu Yunwen.

In the next eight years, Zheng He led large fleets on two more expeditions along the southeast Asian coast and on to the Indian subcontinent. He exchanged lavish gifts with local rulers and displayed superior Chinese technology and

In 1498, Portuguese explorer Vasco da Gama pays court to the ruler of Calicut on India's Malabar Coast. Soon the Portuguese would establish trading forts and decades of warfare would ensue among competing European mercantile powers that followed the Portuguese.

he tapped Cape St. Vincent on the southwest corner of Portugal. "Historic place," he said, glancing at his watch. "That's where Prince Henry had his navigation school. We should be able to see it about now."

Out on deck, the haze had lifted. The stark brown coast of the Portuguese Algarve rose steeply from the sea, which had retained something of the Mediterranean's translucent blue. Whitewashed tourist hotels and vacation bungalows dotted the hillsides of olive and cork oak.

As the coast glided by, I saw a sharper headland of eroded limestone. That was Ponta de Sagres, I realized. For an explorer, this was like completing a pilgrimage.

Today we think of Portugal as a footnote in the modern history of the European continental powers. But in the 14th century, Portugal was one of Europe's first unified nations. This laid the way for a figure I find similar to China's Emperor Zhi Di, Prince Henry the Navigator.

After fighting the Muslim Moors in North Africa as a young man, Henry realized there was no room in the Mediterranean for Portugal to expand: The Arabs controlled its southern shores; the powerful Ottoman Turks held the east. Between them, they monopolized trade with India and China.

But Henry was convinced that tiny Portugal had a global destiny that could be achieved by undertaking epic voyages on the vast oceans. One of his goals was to extend Christianity to Africa and beyond; the sails of all of the ships voyaging under his colors were emblazoned with red crosses. But spreading the faith was not his only goal. Because the direct Mediterranean passage to the treasures of the East was blocked, exploration might discover an alternative. Eventually, finding a practical route around Africa and onward to the spice ports of India would become his life's quest.

In 1419, at the age of 25, Henry established a princely court on the towering limestone headland of Cape Sagres, the southwestern extremity of Portugal. At this outpost he assembled Europe's ablest astronomers, craftsmen skilled in making navigational instruments, cartographers, shipwrights, and sea captains. Henry was certain this elite corps could solve the navigational riddle that restricted Portugal to a narrow strip of land on the Iberian Peninsula: How far south and east did the great continent of Africa extend? Could Portuguese explorers eventually round Africa and make their way to India?

Henry sent one expedition after another down the west coast of Africa. Each one pushed farther south, eventually reaching the great verdant bulge of present-day Sierra Leone. Henry's captains recorded the previously unknown phenomenon of the northeast trade winds. Within a few years, they discovered that the northeast trades could carry a ship almost down to the equator, after which a captain could pick up the steady breezes of the southeast trades. This knowledge proved priceless to later Portuguese seamen attempting to push south down the 3,000-mile length of the lower African coast. By the time of Henry's death in 1460, Portuguese seafarers trained at his school in Sagres were roaming freely down the African coast and out into the open Atlantic.

Many of Henry's ships were caravels designed by the Sagres shipwrights. They combined the deep cargo holds and square sails of medieval European merchantmen with the triangular lateen sails of the eastern Mediterranean. This rig allowed the ship to sail well either before or across the wind, a trait especially important for vessels that regularly traveled both up and downwind in the trades. At about 120 tons, they were

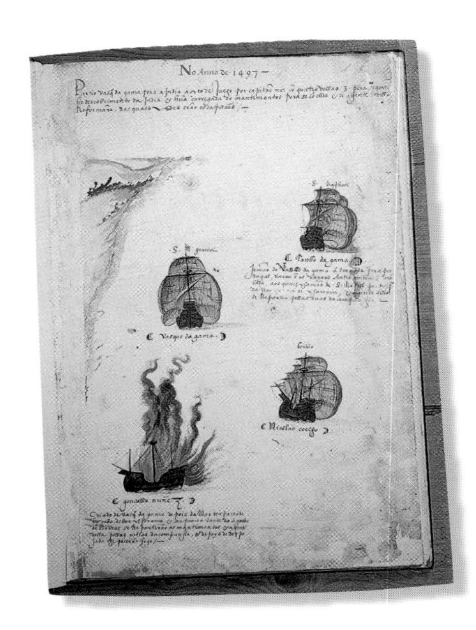

A contemporary view of Vasco da Gama's fleet shows that his flagship *São Gabriel* and the *São Rafael* (upper left and upper right) were deep-hulled *naos*. The smaller caravel *Berrio* (lower right) was an inshore scouting vessel. Da Gama burned the supply ship (lower left) when her holds were empty.

Portuguese explorers under Bartolomew Dias finally rounded the Cape of Good Hope and entered the Indian Ocean in late 1497, almost 30 years after Prince Henry the Navigator's death. The sea route to the riches of Asia was now open. Within 20 years, hundreds of Portuguese merchant sailors undertook this arduous 27,000-mile round-trip voyage.

relatively small, compared to the thousand-ton junks of Zheng He's Treasure Fleet. But caravels were well-found and nimble. In the hands of a competent and adventuresome captain, the vessel was capable of global voyages.

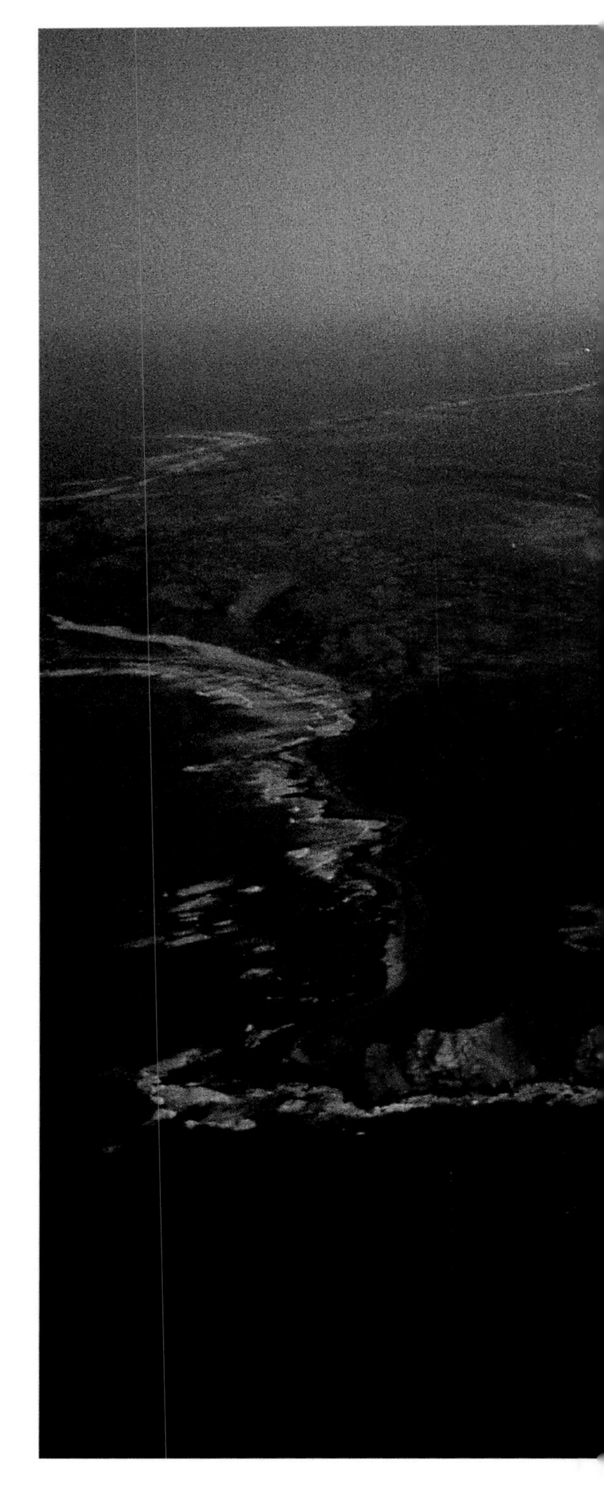

In 1487, Portugal's Prince John ordered a well-trusted captain, Bartolomew Dias, to settle once and for all the question of Africa's southern extent. By Christmas—the Southern Hemisphere summer—his battered flotilla was still moving south under reefed sails in cold squalls, within sight of a low desert coast scoured by hungry breakers. Dias had no option but to continue his passage. He veered off the dangerous shoals, lost sight of land, and plunged ahead into the unknown. Finally the sky cleared long enough for him to take a rough sight of the sun's elevation with his primitive astrolabe and estimate his latitude. It was time once more to steer east to probe the extent of the seemingly endless landmass.

But when Dias and his captains turned and encountered clearing weather, they found the forbidding arid coast had vanished. Now they headed north and land finally appeared, angular, green mountains and rich, grassy valleys. Dias approached a sheltered bay near today's Port Elizabeth, South Africa. He had successfully rounded the southern tip of the vast continent, never sighting the towering headland of Table Mountain at the Cape of Good Hope. After verifying that the green, well-watered coast off their portside ran on uninterrupted to the northeast, Dias was finally convinced that he had entered the Indian Ocean. Prince Henry's water route to the riches of the East had finally been opened.

In 1497, the young nobleman Vasco da Gama was summoned by the royal court to lead a globe-spanning expedition all the way to India. Da Gama prepared a small fleet of mixed caravels and carrack supply vessels for the long voyage. They sailed the curving linked trade wind route down the Atlantic and eventually rounded the Cape of Good Hope in December. A month later, they moved up the East African coast to present-day Mozambique, marking each stage of their passage by erecting inscribed stone pillars called *padraos*.

Most of da Gama's sailors had scurvy; the ships were battered by the stormy passage. But by April 1498, he had pushed on to the coral-rimmed port of Mombasa in today's Kenya. The anchorage was crowded with dhows waiting for the southwest summer monsoon. Despite the suspicion of local Muslim authorities and merchants, da Gama convinced a pilot to lead the Portuguese flotilla across the Indian Ocean to the Malabar coast.

The passage took just over three weeks, all downwind. On the 23rd day, da Gama's lookout spotted rain clouds curling over the Ghats Mountains above the spice port of Calicut. Prince Henry's dream of a water route to India around Africa had been fulfilled.

IV The New World

C.KROHG

Preceding pages: **Sailing west from Greenland with 35 adventurous explorers in about 1000 A.D., Leif Eriksson sights the rock-bound coast of Baffin Island and makes landfall on a new continent. The Vikings turned south and eventually established an outpost in Newfoundland.**

My boots crunched on the snowy cinders as I followed my colleague, geologist Haraldur Sigurdsson, up the western shoulder of Hekla, one of Iceland's major volcanoes. Above us a wide crater steamed, a mute reminder of the last eruption in 1980, nine years ago. Haraldur pointed south toward the blue-gray Atlantic.

"You can see Heimaey quite well," he said. "But Surtsey's cone is much smaller."

The two islands lay off Iceland's coast, dark mounds against the pale northern sky. Having only risen from the sea in 1963, Surtsey's black lava was devoid of vegetation, indicating its recent eruption from the ocean floor and providing stark testimony that Iceland stood astride one of the world's most active tectonic zones.

The island nation's western third lay on the North American plate; the eastern portion was connected to the Eurasian plate. Beneath us, a dome of molten magma ballooned up to fill the long cracks formed as the two plates pulled relentlessly apart. Iceland was an area of intense volcanism, a rising lump on the backbone of the Mid-Atlantic Ridge, which snaked south to the Azores, on to Ascension Island, then swung east to enter the Indian Ocean.

To our right, lumpy ridges of green pasture, interspersed with tan hay stubble, rolled toward the capital, Reykjavik, 70 miles to the west. Off to the left, we saw no grazing land or fields, only bleak lava flows, scattered pewter lakes, and steaming cinder cones. On the eastern horizon, the creamy Vatnajokull glacier rose to the summit of another volcano, Grimsvotn.

"Now I know why the tourist office calls this 'The Land of Fire and Ice,'" I told Haraldur.

He laughed, then nodded toward Hekla's main crater. "The Vikings named this mountain the 'Gate of Hell.'"

Haraldur explained that Hekla had played an important role in the pagan mythology of the island's earliest settlers, daring Norse seafarers who had ventured forth from Scandinavia before the year 900 A.D. to settle this forbidding land. Repeated volcanic eruptions, with terrifying blasts of incandescent lava and roiling clouds of poisonous smoke, inspired the apocalyptic Viking poem "Ragnarok," "The Doom of the Gods." In this epic myth, demonic giants rise from the sea (newly formed volcanic seamounts like Surtsey), the sky darkens (obscured by the ejected ash of major eruptions), and the land sinks beneath the icy waves, only to rise again (a fair description of coastal upheavals on fault lines). In many ways, I realized, as Haraldur amused me by reciting rhythmic stanzas of Old Norse from the poem, the Viking explorers were accurately recording the phenomenon of Iceland's tectonic volcanism—within their culture's metaphorical limits—just as the early Greeks had described Crimean gold panning using sheep skins in the myth of Jason and the Golden Fleece.

As Haraldur chanted the ancient stanzas, my thoughts wandered back over the centuries to one of history's most heroic periods of exploration, the ocean-spanning voyages of the Vikings. I could almost see a fleet of sleekly planked *skuta* roundships, their woven wool square sails tautly filled with the brisk northeast wind, escorted by a pair of lean *snekkja* longships, each oarsman's position guarded by a burnished shield. It was in open-decked vessels such as these, none longer than a hundred feet, that the Vikings transformed the stormy waters of the far North Atlantic

I climb from an active fissure in Iceland's Krafla volcano during a 1988 expedition. Iceland sits astride the Mid-Atlantic Ridge, the island divided between the North American and Eurasian Plates. Volcanism produces lava flows and geysers.

into their private lake, colonizing Iceland and Greenland, and eventually making the first European settlements on the continent of North America.

The Vikings were descendants of Scandinavian tribal people, who by the ninth century A.D. had become the most skilled and audacious seafarers in Europe. They were also the region's most feared warriors, repeatedly plundering the hapless principalities of the British Isles and later planting colonies there and along the mouth of France's Seine River and Normandy. In the east, the feared Norsemen rowed their longships through the shallow waters of the Baltic coast, slashing aside all who resisted. The Vikings even penetrated the ancient heart of Russia down the Eurasian river systems and descended as far as Kiev in modern Ukraine.

But the Vikings who sailed west into the open Atlantic, while retaining their military prowess, were primarily exploring colonizers, not raiders who had extensive settlements in the British isles. They were following local legends of mysterious, potentially fertile lands that lay beyond the horizon of the great ocean.

The most alluring of these Celtic myths concern the Irish monk Saint Brendan, who, according to the medieval account *Navigatio Sancti Brendani*, traveled the western ocean for seven years (566 to 573 A.D.) on a missionary voyage that bore heroic

echoes of Homer's *Odyssey*. Brendan and 17 colleagues sailed in an ox-hide *curragh*. They encountered fur-clad pygmies, who might have been Eskimos, and a terrifying region where the seawater was a "curdled mass," which could have been the tangled yellow weed in the Sargasso Sea. After sailing north along the shore of a "large continent" (perhaps North America), Brendan's party saw a mountain of "crystal," possibly their description of an iceberg. Although never confirmed, the quasi-mythical voyage of Saint Brendan contained enough intriguing description to have whetted the appetite of adventure-hungry Viking voyagers.

Whether their motives were pure adventure, population pressure in the crowded fjords of their Scandinavian homeland, or banishment for major crime—as was the case for many Vikings—the Norsemen took advantage of a cyclical warming trend in the northern hemisphere beginning in the ninth century. Vikings from Norway made the first settlement on Iceland in 874; the glaciers had retreated from the coastal valleys, leaving plush grass for grazing. The close-in waters teemed with cod.

But the settlers' Icelandic Commonwealth was not a utopia, despite the relatively warm weather and the productive fishing grounds. The Vikings had carried with them from Scandinavia the ancient tradition of the blood feud, in which a murder in one clan had to be repaid with another in the offending clan. Harshly autocratic fathers ruled their households; heavy labor was performed by slaves.

One of these violent Norsemen was named Thorvald Asvaldsson; he had been banished from Norway for manslaughter. With his large family and retinue of slaves, Thorvald sailed the 700-mile passage to Iceland only to find that the best pastures and farming plots had been taken. He was forced to settle on exposed land in the western hills. There his family scratched out a living. Thorvald's son, Erik the Red, fell into another blood feud and was banished in 980. He sailed farther west with his exiled household toward the snowy mountains he had first seen in his youth. But

Fresh lava flows in the early 1970s devoured Icelandic coastal villages and the fishing port on the offshore island of Heimaey. Like their rugged Norse ancestors, today's Icelanders must seek a precarious harmony with their restless island. Volcanism provides limitless geothermal energy, but unpredictable advancing lava flows can destroy a town overnight.

as Erik's boat approached the towering glaciers, it became grimly obvious that this coast offered little promise as a place of comfortable exile, or even a refuge for survival. He cruised past ice-choked fjords, landed on the more sheltered southwest coast, and built a settlement called *Brattahlid.*

Ever optimistic, Erik saw the possibility of a large settlement based on farming, grazing, and fishing. In mid-summer 985, Erik returned to Iceland to champion settlements in "Greenland," the deceptively hopeful name he had given this immense glacier-bound island. The next year he led a party of 500 colonists aboard 14 ships, carrying several years' provisions, as well as horses and cattle. While the climate remained favorable and the settlers could harvest their hay and grain, the Greenlanders lived well.

In 1976, explorer Timothy Severin (center) built the *Brendan*, a replica of a medieval ox hide curragh. Using flax thread, he sewed the hide to a frame of ash laths. The boat retraced the transatlantic voyage of the legendary Irish monk, St. Brendan.

But the Vikings' spirit was still restless. Erik's second of three sons, Leif, who had been born in Iceland around 980, had spent his youth building up the Greenland colony. But he was also a skilled and daring seafarer. He sailed directly from Greenland to Norway around 1000 A.D. There King Olaf I converted the young Viking to Christianity and dispatched him back to Greenland to spread the Gospel. Leif also heard accounts of a Bjarni Herjolfsson, who had sighted mysterious and enticing lands to the west. Herjolfsson, returning to Greenland from Iceland, had been blown off course by violent gales. After sailing for days to escape the storm, he sighted a long, well-timbered coast as he tacked his way back north to the latitude of Greenland. But Herjolfsson was a trader, not an explorer; he hadn't risked his cargo by venturing ashore.

Leif Eriksson quizzed the older Viking closely about the route he had followed, then struck a bargain to purchase his boat, a well-found skuta roundship capable of carrying a large crew, provisions, and, equally important, a cargo of timber on the return voyage.

Eriksson and 35 men sailed almost due west from Greenland, hoping to retrace Herjolfsson's course. Their first encounter with the North American continent was probably the glacier-scoured edge of Baffin Island. He called this desolate region *Helluland* (the coast of rocky slabs). There was nothing here to lure the Vikings ashore. Eriksson turned south. After several days sailing, they entered a wooded cove, which was probably halfway down the coast of today's Labrador. Although the pines were dense, they seemed stunted compared to the trees Herjolfsson had described. Eriksson's party continued farther south, finally reaching a well-forested headland that stood on a peninsula. A narrow river tumbled from an inland lake, providing ample fresh water. The beaches and tidal rocks teemed with shellfish. The Vikings decided to make a winter camp.

Following the building habits acquired in timber poor Greenland, Eriksson's men constructed sod-roofed houses, surrounded by a tall woven stockade of saplings surmounted by sharpened stakes. This first European settlement in North America was probably at present-day L'Anse aux Meadows on the extreme northeastern end of Newfoundland. They called the site *Leifrsbudir* (Leif's Huts). The local water was rich with migrating fish, and the Norsemen caught an ample supply to be dried for the winter.

As the Vikings explored the area in small parties, one of their members brought

back the exciting news that the woods near the lake were thick with wild grapes, a sign that this peninsula was blessed with a mild climate. Eriksson named the region Vinland.

In the spring of 1002, Eriksson returned to Greenland with a cargo of timber and dried grapes. While making plans for a larger, colonial expedition to Vinland, Leif Eriksson's father died, leaving his son in charge of the Greenland settlement. Future exploration of Vinland fell to Leif's brother, Thorvald. He returned to the Leifrsbudir settlement in 1004. The next spring he led a party farther north to *Markland* (Forest Land) and was killed in an encounter with hostile indigenous coastal people the Vikings called *Skraelings*, whom Leif Eriksson's initial party had not encountered.

Several years passed before the Vikings again attempted to exploit Vinland. In 1010, an expedition of 160 men and several women sailed aboard three ships to build up the Leifrsbudir enclave. At least one Viking child was born at the settlement, which the Norsemen probably used as a base camp from which to explore a much wider area. They cut timber and traded metal goods for furs with the Skraelings during periods of uneasy truce.

Little is known of the Vikings' continued contact with North America after this, although the Norse artifacts that archaeologists have unearthed at L'Anse Aux Meadows indicate the settlement lasted several years and was self-sufficient. But the increasingly hostile native people probably dampened the questing zeal of even the adventurous Norsemen.

George Molony pulls the *Brendan* ashore on Peckford Island, Newfoundland, on June 26, 1977. The craft held up well during the stormy, five-week passage. Severin chose his route by matching descriptions in the medieval account of the sixth century Irish monk-explorer's voyage with prevailing winds and currents. The crossing demonstrated that the Irish explorer could have reached North America 500 years before the Norsemen.

As a WHOI graduate student in September 1967, I went on my first major scientific expedition, a seismic survey of the sedimentary rocks of the Atlantic continental rise aboard the Institution's research ship *Chain*. To probe those rock layers 8,000 feet below, we used a towed high-energy sound wave generator, the "sparker," which was virtually a lightning-and-thunder machine. Generators charged huge capacitators with more than 100,000 joules of static electric energy then transmitted that tremendous pulse to the sparker, which fired it into the water from the tip of an electrode serving as a giant spark plug.

The resulting explosion was bright enough for orbiting astronauts to see at night. The shock wave shook the 1,800-ton vessel along her entire 200-foot hull. These

relentless thunderclaps recurred every 20 seconds. And the smacking echo returning from the bottom bedrock was almost as disturbing.

"That was a good one," my colleague Al Uchupi muttered between puffs on his contraband Cuban cigar after an especially loud sparker blast on the second night out.

"I guess I'll get used to it," I answered, glancing aft.

The lab's bulkheads twanged and the steel deck beneath my boat shoes shuddered, as if the *Chain* had just dropped a depth charge—or been hit by lightning. But outside the open porthole the sea was burnished with silver moonlight, and fair weather clouds drifted past.

After Leif Eriksson (foreground) led the first European expedition to the New World, the Norsemen returned over the next ten years, exploring beyond their outpost at L'Anse Aux Meadows in Newfoundland. But conflict with Native Americans eventually drove the Vikings back to Greenland. Had the Vikings reached a more hospitable environment farther south, the history of the Western world would have been forever altered.

To write legibly in the data logbooks in the science lab, perched high on the ship's superstructure, I had to wedge myself at my workstation. Al and I stood the graveyard watch, midnight to 4:00 a.m., a challenge that I found both exhausting and exhilarating. The expedition leaders, including WHOI's Professor K.O. Emery, trusted us to gather precious data, the lifeblood of science. This gave me a sense of membership on a big-league expedition for the first time. I was determined to meet my responsibilities.

The lab held dozens of instruments, which we had to constantly monitor. They whirred incessantly, spewing endless trains of seismic tracing. *Chain*'s recording echo sounder also dumped a long ribbon of paper, measuring the depth, while other instruments measured the Earth's magnetic and gravitational fields.

We had to annotate and log all this data every five minutes. In our few spare moments, we changed the magnetic tapes backing up the instruments' paper traces. Several times each watch, Al and I performed a routine as we quickly swapped about two dozen logbooks to be annotated.

I soon discovered the Spartan reality behind my romantic notions of science at sea. Living on a vessel like *Chain*, which had been designed as a seagoing salvage tug during World War II, was hardly comfortable. As the most junior member of the WHOI party, I drew a top berth in the crowded foc's'le, my bunk only 18 inches below the main deck. I was in deep sleep on the second morning of the cruise, despite the incessant blast of the sparker. Then bo'sun Jerry Carter cut loose with a pneumatic rust chipper on the deck plates just above my head. I bolted upright and slammed my skull into the steel, gashing my forehead. I stood my watches, the wound bandaged, and my head throbbing every time the sparker exploded.

Then things got considerably worse.

On September 17, Hurricane Dora churned up the warm waters of Gulf Stream, straight toward us. The hurricane's eye was well out to sea, but *Chain* was caught between the storm and the dangerous shoals of Cape Hatteras. In 1967 weather satellites were incapable of providing us a clear picture of Dora's track. In only hours, a muggy, overcast day with a slowly mounting easterly chop had deteriorated into a full gale. Incredibly, no matter which way we headed, the rumbling mass of squall clouds and pounding, 30-foot waves also seemed to change direction, as if the storm anticipated our new heading.

Some wag in the crew painted a square of plywood with an arrow and the word DORA and fixed it firmly on the bow rail, to show defiantly that we were ready for the storm that stubbornly dogged us.

Now survival of the ship, not science, had become the crew's responsibility. They struggled under dangerous conditions to winch in the sparker and our towed hydrophone arrays, then to secure deck equipment with double lashings of steel cable. Still, the big blue derrick swayed precariously as the wind blasted *Chain* with 80-knot gusts, and solid green sea burst across her deck.

The skipper kept the bow into the sea and wind, and maintained enough speed so that the huge waves would not roll *Chain* broadside in the deep troughs. After sealing watertight hatches and portholes, we waited. Despite the seals, water found its way into the ship to slosh through passageways and mix with the vomit of the seasick.

On the second morning, I had to have fresh air and cautiously made my way up to a sheltered porthole that the crew had allowed us to open for brief periods. Outside, the storm was abating. But the sky was still dark gray, hunkered low, the wind howling. The sea remained a confused, heaving mass, with nasty cross-waves slashing atop huge surging dirty green swells. For two days we had traveled southeast, away from land, beyond the warm Gulf Stream water that fueled the hurricane.

Ahead lay the Bahamas, the Caribbean. As I watched the angry remnants of Hurricane Dora, which had disrupted my first major seagoing expedition, I pictured the frail ships of the first European explorers to reach these waters. Christopher Columbus's tiny caravels each weighed less than a tenth of *Chain*'s displacement. Yet his small fleet had entered the treacherous reefs of the Caribbean at the height of the hurricane season. Could they have survived a storm such as Dora? Of course not.

In this fanciful 19th century impression of Columbus sailing from Palos on August 3, 1492, his patrons, Queen Isabella and King Ferdinand bid him farewell. In fact, they did not. But they did treat him royally on his return from the first successful voyage to what he thought was Asia.

Watching the angry sea outside the porthole, I wondered how history would have been affected if Columbus had encountered such a storm.

When I was in grade school, the name Christopher Columbus was synonymous with exploration and discovery. But a lot of what I learned about him back then is incorrect. Columbus did not "prove" the world was round by sailing west across the Atlantic in search of Asia; competent Renaissance navigators realized Earth was a spherical planet.

Nevertheless, he remains history's preeminent explorer. Had he not persisted in overcoming nearly insurmountable obstacles, Renaissance Europe's discovery of the New World might have been delayed for decades or even centuries.

Born into a Genoese family, he became an experienced seaman by 25 then joined his younger brother Bartholomew in Lisbon, where they worked as cartographers. Christopher studied classical geography and Marco Polo's account of his overland journey to China, melding Ptolemy's underestimate of the Earth's circumference with the Italian's gross overestimation of the Asian landmass's eastern extent. A concept took root in that Lisbon map shop: The Ocean Sea was not as vast as many thought, but just over 1,000 leagues (about 3,000 miles) wide. And Marco Polo had described the wealthy islands of Cipangu (Japan) as lying well to the east of mainland Asia. Their latitude was known; reaching these outlying islands and Asia itself by sailing west seemed feasible.

In 1484, Columbus tried to interest Portugal's King John II in his "Enterprise of the Indies," the idea of sailing west across the Ocean Sea to Asia. The Portuguese court, however, still backed its around-Africa route to India, which ran exactly opposite to Columbus's scheme. And the royal cartographers saw that Columbus had underestimated the distance westward across the Ocean Sea to Asia, which they had correctly calculated to be at least 3,000 leagues (around 10,000 miles). Columbus received no support from the Portuguese throne.

Columbus received harsher treatment in 1500 when he was dispatched to Spain in chains following his third troubled voyage to the New World, during which he hanged rebellious Spanish settlers in the Caribbean.

But Spain's King Ferdinand and Queen Isabella appointed a commission to study his proposal. Again, geographers found serious flaws in his concept. Columbus was close to abandoning his initiative, but found allies among Franciscans eager to establish missions in Asia who helped him obtain another audience with Queen Isabella. Her advisers reversed themselves, urging the queen to support his enterprise. Columbus was named Admiral of the Ocean Sea, and was to receive 10 percent of the venture's profits.

To prepare for the epic voyage, Columbus worked with a Spanish seafaring family, the Pinzons. He selected the *Santa Maria*, a square-rigged carrack of about 100 tons and two smaller caravels, the *Niña* and the *Pinta*. On the advice of Martin Alonzo and Vicente Yanez Pinzon, who would command the caravels, Columbus chose 90 experienced seamen, not the former convicts mentioned in my schoolbooks. For his chief pilot, Columbus turned to Juan de la Cosa, an able navigator. Columbus carefully stocked his tiny flotilla with ample provisions, armaments, and trading goods.

On the morning of Friday, August 3, 1492—well into what we now recognize as the hurricane season—the three ships left the Spanish port of Palos de Frontera on the ebbing tide and made for the open sea. Columbus set a course directly for the Canary Islands, convinced that Japan lay at the same latitude. Taking advantage of the prevailing northeast trade winds, all Columbus had to do was sail "down the latitude" for about a thousand leagues before he made his landfall.

Beyond the Canaries, they encountered frustrating calms for several days. Finally, the trade winds exerted their influence and the small flotilla rolled peacefully along, day and night. Yet the farther west they sailed, the more uneasy the crew became. They were Spanish sailors unaccustomed to following the northeast trades so far from land, unlike their Portuguese counterparts. Panic was spreading among the crew: If the Admiral's theories were correct, they should have reached Japan by now.

Then the trade winds failed for ten days and the ships drifted aimlessly on the swell, the slack sails rattling in the heat. Rafts of bizarre yellow sargassum weed dragged along the vessels' sides, fouling their rudders. Now the crews' mutinous grumblings became open. On October 10, aboard the *Santa Maria*, a delegation of seamen brusquely demanded that Columbus return to Spain. He signaled the captains of the other two vessels to join him for a conference. Martin Alonzo Pinzon suggested the fleet had drifted north of its intended latitude; by turning southwest, they should quickly find the islands of Asia.

Columbus ordered the course change. Within 24 hours, they sighted palm fronds and branches floating in the sea. Obviously, they were nearing land. Talk of returning to Europe ended. Four hours after sunset on October 11, 1492, Columbus stood on the quarterdeck of the *Santa Maria* staring intently toward the southwest. He told the helmsman to ease the ship slightly downwind. Then Columbus caught a momentary glimpse of a distant orange spark. Was it a fire ashore, or simply an illusion?

Three hours later, at 2:00 a.m. on October 12, a lookout in the rigging of the *Pinta*, Rodrigo de Triana, cupped his hands and called hoarsely, "*Tierra!*" He had seen a pale sandy shore, unmistakably vivid in the moonlit night.

With daylight, Columbus's vessels anchored in a sheltered bay near a palm-lined beach. Columbus called his captains together and proclaimed the island, which he christened San Salvador, a possession of Spain.

Columbus led three more expeditions, opening the door of what would quickly become the world's richest empire, Spain, always hopeful that the Spice Islands lay beyond the next green cape. On his last voyage, after shipwrecks and a year's marooning, Columbus sailed from the island of Hispaniola for Cadiz in September 1504, his health broken. It was doubtful he could muster the energy or the patronage needed to explore again. Still, he followed the Spanish royal court to Valladolid in April 1506, hoping to secure sponsorship for one more voyage of discovery. He died in that ancient city on May 20 at age 55, still believing he had pioneered the sea route to Asia.

During the years I lived on Cape Cod, there was a stop I loved to make about once a week in the village of East Sandwich. The fly-fishing shop and the antique map store were adjacent, my idea of combining pleasure with pleasure. As my informal study of maritime history progressed, I found myself more engrossed in the dusty volumes of parchment maps and spent less time fiddling with the latest expensive creations of pheasant feather and fine copper wire meant to lure even the most spooked brook trout out of his haunt.

The older the maps, the less accurate they were, and the more absorbed I grew in their history. The woman who ran the store knew I couldn't afford her really valuable items, but she kindly let me study them, provided I wore curator's cotton gloves.

Some winter afternoons on the drive home from Woods Hole, I'd find I had spent three hours poring over 18th century British Admiralty charts of the Windward Islands or crumbling Portuguese colonial maps of Amazonia. It wasn't so much that these precious documents were accurate measurements of the Earth's surface—although many were amazingly precise, given the surveying instruments their makers had used—but that they were exploration made tangible. The succession of old maps and charts represented the widening human understanding of our planet.

That quiet little antique shop on a Cape Cod side road was a substantive link back to the great Age of Exploration.

As news of Columbus's voyages spread, European geographers clambered for details of the lands that lay in the previously empty blue longitudes on their rudimentary globes. Determining just how far "Asia" extended east into the Ocean Sea became a priority in the early 16th century. It was another Italian, Amerigo Vespucci, who solved this mystery. And, in the process, he gave his name to both continents in what became known as the New World.

Son of a rich Florentine merchant family, Vespucci was employed in maritime affairs in Seville when Columbus returned from his first voyage in 1493. Vespucci became a close friend of the explorer and his lieutenants, including an energetic young Spaniard named Alonso de Ojeda.

When Ojeda invited Vespucci to join an expedition of discovery in 1499, the Florentine merchant-cum-navigator readily agreed and gained the right to explore on his own. His objective was to determine the extent of the "Chinese" coast that lay south of the islands Columbus had earlier discovered.

Vespucci, in command of four small ships, steered due downwind on the northeast trades to landfall on present-day Brazil near the equator. He then diligently followed the jungle coast to the northwest, pausing at the Amazon estuary to note the incredible volume of fresh water that stained the blue tropical sea a muddy brown far offshore. Vespucci recognized that such an immense river had to drain a continent. And when he reached the mouth of the Orinoco, in what is today Venezuela, several weeks later, he observed a similar phenomenon (as had Columbus on his third voyage).

He was convinced he had found an eastern promontory of continental Asia. If he had sailed farther south, he reasoned, he might have entered the Indian Ocean from the east and reached the Spice Islands or the Malabar Coast.

But back in Spain, Vespucci was unable to interest the royal court in this line of exploration. So, reversing the process Columbus had followed ten years earlier, Vespucci turned to the Portuguese. King Manuel I was taken by Vespucci's proposal, especially because much of the coastline he intended to explore might lie on the Portuguese side of the Papal Line of Demarcation (A Vatican council had divided non-Christian territory across the Ocean Sea between Spain and Portugal).

Vespucci's fleet, their sails emblazoned with Prince Henry's red cross, left the Tagus River in May 1501. They slowly explored their way down the coast of Brazil, probing each promising bay and river for a possible sea route around the continent to the Indian Ocean. But as they crept farther south, the shoreline continued unbroken. The contacts they had with native people gave no evidence this land was a part of Asia. These simple Indians knew nothing of spices or the Great Khan. Still, Vespucci was determined to press on as far as possible, even though he was now well below the equator and had found no navigable passage east. Finally, after venturing almost down to latitude 50 south on a bleak, windswept coast, Vespucci turned north for the voyage back to Portugal.

He was now convinced that the long coastline he had explored was part of a newly discovered continent, not a peninsula of Asia. As he worked on his charts during the return journey, Vespucci took his quill to the vellum and wrote the words *Mundus Novus* (New World).

In this 16th century allegorical painting, Amerigo Vespucci (left) holding pennant and astrolabe pays court on "America." In the background, her tribe roasts a human leg on a spit, symbolizing the disdain with which many Renaissance explorers held native people.

His published journal of this expedition was a sensation in literate Europe and was quickly printed in many editions and languages. Soon geographers were describing not just the new western continent Vespucci had explored, but all the territories of the New World as "America."

In 1981 I joined the Scripps Institute of Oceanography research ship *Melville* in Valparaiso, Chile, and spent two uninterrupted months out at sea investigating hydrothermal vents on the East Pacific Rise at 20°south latitude.

This extremely active tectonic zone is the world's fastest-spreading ocean floor, where the plates are separating at 20 centimeters a year. As on the Galápagos Rift, our primitive but trusty old "Dope on a Rope" camera sled *ANGUS* photographed colonies of platter-sized white clams, bizarre red-mawed giant tube worms, and other assorted bizarre creatures living without the benefit of a photosynthesis-based food chain around the sulfide-rich deep-sea hot springs.

The confirmation that the Galápagos Rift menagerie existed elsewhere under similar conditions was not the most profound revelation I took from the long expedition. Although I had spent several weeks at sea before, I'd never felt such a sense of utter isolation. Our ship cruised a vast area of the Pacific almost devoid of islands, and we certainly never approached any. Week after week, all we saw was the endless blue rim of the tropical ocean horizon, occasionally *continued on page 116*

Pirates of the Spanish Main, the Caribbean islands and coasts of Central and South America, preyed on the region's rich commerce for centuries, following soon after the plundering conquistadors. In the 16th and 17th centuries, buccaneers commanded large fleets and thousands of men. No ship or port was safe from their ruthless attack.

Pirates of the Spanish Main

By the mid-1500s, the Spanish used the term Tierra Firma to differentiate the coasts of Central and South America from the islands of the Caribbean. The English translated this term as "the Spanish Main," and applied it to the entire region. Whatever words were used, the area became synonymous with unimaginable wealth.

The original uneven flow of gold from the Caribbean became a torrent after conquistadors plundered the ancient Aztec and Inca empires. The first treasure Hernán Cortés seized from Moctezuma in 1520 consisted of a trove of Aztec royal jewelry and palace ornaments, which had been fashioned from gems, nuggets, and gold dust panned over generations in streams. To their dismay, however, the Spanish captains found no native gold mines in Mexico ready to loot. But the conquering Europeans did discover rich silver deposits, which they exploited ruthlessly using local people as slave laborers. This cruel sequence was repeated when Francisco Pizarro overwhelmed the Inca of Peru in the 1530s. When the colony of Cartagena was established on the coast of present-day Colombia, a lavish new source of gold and gems was discovered.

In 1535, the first of many Spanish treasure fleets assembled in the Caribbean for the voyage to Seville. Typically, these armed merchantmen carried gold and silver ingots, pearls, emeralds, and gold jewelry looted from Aztec and Inca tombs. Later, as the Spanish mining enterprises of the New World matured, the Crown established mints in Mexico and Peru that produced the famous "pieces of eight," which became the standard currency of cash-starved Europe and—under the name of "the Spanish milled dollar"—remained legal tender in the United States until just before the Civil War. The outpouring of wealth from the New World—added to the swelling flow of riches from Asia—fueled mercantilism in Europe. This new affluence spread among the privileged classes who helped stimulate the subsequent Enlightenment, which in turn created the intellectual environment necessary for the industrial revolution of the 19th century.

It was inevitable that the flood of treasure from the Spanish Main attracted both privateers—warships bearing a "letter of marque" from a monarch granting the captain the right to prey on enemy commerce—and independent outlaw pirates who answered to no authority. Piracy, of course, was not unique to the Spanish Main. Homer's Odysseus was said to wantonly plunder ships and coastal towns; a thousand years later, the Roman leader Pompey the Great had to mount a major expedition to clear the Mediterranean of brazen pirates who threatened to strangle trade. But at no point in history were pirates and privateers so active as in the 16th and 17th centuries along the Spanish Main.

And one of the boldest was an Elizabethan Englishman the Spanish called "El Draque" (The Dragon), Sir Francis Drake. Born into a poor Protestant farming family in 1540, Drake went to sea at age 13 and became an accomplished sailor aboard a merchant vessel in the storm-swept North Sea. As a young man he joined two English commercial voyages to the Caribbean, sailing in defiance of the Spanish claim to sovereignty over the region.

On the second voyage in 1568, Drake commanded the *Judith*, and was treacherously attacked by the Spanish on the Mexican coast. Many of Drake's sailors were killed. But he successfully brought the damaged vessel home, vowing to reap revenge against the "Dons" and their Catholic monarch, Philip II.

Queen Elizabeth I, a bitter enemy of the Spanish throne, was taken by Drake's bravery and seamanship. Calling him "mine own pyrate," she granted Drake a letter of marque: the right to cruise the Spanish Main as a privateer. He sailed in 1572, commanding two vessels barely larger than yachts, the *Pasha* and the *Swan*. Drake's goal was bold though, plundering Cartagena and Nombre de Dios on the Caribbean coast of Panama, vital entrepôts for treasure from across the Spanish Main. Although wounded, Drake pushed inland to stand on a mountain and study the blue expanse of the Pacific beyond the isthmus. There he prayed for the chance to sail that sea under the flag of England.

Drake returned home a wealthy man, but was not officially rewarded because of a periodic truce between England and Spain. When the war resumed in 1577, Drake was sent on his most audacious royal commission, a voyage south and west of the American continent claimed by Spain and Portugal, in search of new lands and possible prizes. After passing through the Strait of Magellan in the fierce southern winter, one of Drake's ships sank, another returned to England, and only his flagship, the *Golden Hind*, remained. Nevertheless, Drake plundered the Spanish port of Valparaiso and seized a treasure galleon, the *Cacafuego*.

Armed with captured Spanish charts, Drake sailed as far as Vancouver Island in North America, then claimed the coast of California for the English Crown. He continued across the Pacific, eventually reaching the Philippines, where he ambushed Spanish ships just as they had waylaid the *Judith* years before. In the Moluccas, he took aboard a cargo of spices, even though the ship was already laden with gold and silver. Drake reached Plymouth in September 1580, having circumnavigated the globe and captured an immense treasure. The Queen knighted him aboard his ship.

For the next ten years, Drake served as both a naval commander—he was the English vice-admiral in the 1588 victory against the Spanish Armada—and a privateer, returning to the Spanish Main several times, where his presence spread terror among the Spanish. He died of dysentery off the coast of Panama in 1595.

Drake's exploits ignited a surge of piracy and privateering along the Spanish Main. Some of the boldest outlaws were "the Brethren of the Coast," who became known by the name forever associated with their bloody profession: buccaneers. In the early 1600s, the northern half of the island of Hispaniola, an uninhabited stretch of savanna, became a refuge of fugitive bond-servants, deserters from ships, and escaped convicts. Out of necessity, they formed an alliance, surviving by hunting feral livestock from the abandoned Spanish colonies of the previous century. The French term boucaniers derived from the wooden smoking grills these desperate men used to prepare jerky, which they bartered to passing ships for necessities such as gunpowder, shot, and the occasional keg of rum. Dressed in rags, they rarely bathed and smeared themselves with animal fat to repel mosquitoes and ticks.

When the Spanish authorities eradicated the wild cattle and hogs, the buccaneers took to the sea in their dugout canoes and began to attack the vessels with which they had once traded. The pirates shifted base to the small island of Tortuga where they formed an organized federation. Open boats gave way to sloops equipped with swivel guns; the sloops were supplanted by captured armed merchantmen.

Henry Morgan (left), a rapacious buccaneer, sacked the rich Spanish entrepôt port of Panama on the Pacific coast of the isthmus in 1670. He held wealthy Spaniards like Don Pasquale (right) for ransom, extracting their last ounce of gold and gems.

Soon the Tortuga buccaneers were combing the Caribbean in organized bands, raising havoc with the Spanish treasure fleets.

With titular control of Tortuga disputed among the English, French, and Spanish, the main buccaneer fleet settled in Port Royal, a fortified harbor on the south coast of Jamaica. Although now officially an English colony, the island's only real defense rested with the buccaneers. Since their main prey was the Spanish, England was happy to grant the buccaneer federation this safe haven. This arrangement suited both sides: The English Crown's share of the buccaneers' spoils was so lucrative that London considered building a mint in Port Royal. But this proved unnecessary as the pieces of eight the buccaneers delivered to the Crown were so plentiful, additional coinage was not needed.

The heyday of the buccaneers and the darkest days for the Spanish in the Caribbean came with the arrival of a Welsh seaman named Henry Morgan. As a young Navy officer, he served in the expedition that captured Jamaica from the Spanish in 1655. Morgan knew the risks of attacking Spanish treasure convoys, protected by well-armed escorts. But boldness was, indeed, the key. His first victim was Puerto Principe on Cuba, which he savagely sacked. In an even more audacious move in 1668, Morgan led a brilliant assault against the fortress of Porto Bello on the Caribbean coast of Panama. The pirates captured the citadel, tortured their prisoners into revealing where they had hidden their valuables, then held the city ransom for 100,000 pieces of eight. Morgan's buccaneer force next ventured into Lake Maracaibo, deep within the Gulf of Venezuela. His ships returned to Jamaica laden with gold, silver, and emeralds.

By now, every governor on the Spanish Main either fortified his treasury, or removed the precious contents safely inland. But the Spanish authorities did not reckon on Morgan's daring or the ruthlessness of his buccaneers. In August 1670, he assembled 36 ships and over 2,000 pirates and prepared to attack Panama on the Pacific coast of the isthmus.

After an advance party captured the fortress guarding the Chagres River, Morgan brought up his main fleet and put his men ashore. He divided 1,200 buccaneers among longboats and canoes and began the journey upriver into the malarial interior of Panama. The men were heavily laden with cutlasses, firearms, shot, and powder, but carried little food. After ten days on the river, and climbing steep jungle trails, the buccaneers stood on the ridge above the Pacific port of Panama. Thick lines of Spanish troops were arrayed around the city's rough-stone walls. The Spanish cavalry attacked, but were bogged down in the mud. After the enemy forces retreated within the walls, Morgan's famished and exhausted buccaneers assaulted the citadel with brutal frenzy and overwhelmed the Spanish.

For the next three weeks, Morgan's men methodically sacked the city, burning the homes of prosperous merchants when they had been pillaged. Wealthy prisoners were tortured to reveal where they had cached their family's treasure. Other prisoners were held for ransom, so that every possible gold ducat, silver piece-of-eight, pearl, or emerald was wrung from the hapless captives.

When the buccaneers finally marched away on February 24, 1671, Panama was a smoking ruin. Morgan conducted the traditional division of the spoils as the surviving party assembled on the Caribbean coast. He cheated his men of their rightful due, then quickly departed, his own ship crammed with loot. So much for the adage, "honor among thieves."

Pieces of eight, also called the Spanish-milled dollar, began flowing from the Caribbean in the 16th century. The bounty treasure fueled mercantilism in cash-starved Europe, and the coins remained legal tender in the United States until the Civil War.

swept by the black curtains of squalls. Sometimes there were dolphins or small pilot whales riding our bow wave, but no birds ventured this far from land.

I enjoyed the sense of solitude, the time to read between watches in the control van and lab. The ship had a good library and an ample collection of videocassettes. And the Happy Hour beer on the fantail with the trade wind blowing and the Southern Cross winking on like clockwork was always pleasant. After seven weeks, however, as we finished our scientific work and turned west toward Nuku Hiva in the Marquesas Islands, I found myself almost hungering for sight of land.

When those impossibly green volcanic seamounts rose ahead on May 23, some of us lining the ship's rail cheered. Now, I thought, I knew what it was like to have been at sea. I believed I could finally grasp in a visceral way that the Pacific comprised one third of the planet's surface.

The beautiful Bay of Virgins, Fatu Hiva, Marquesas Islands is a green oasis in a blue desert. The Pacific Ocean is so vast that in centuries past mariners, such as Ferdinand Magellan and those who followed his pioneering circumnavigation, could travel for months without sighting land. Inevitably, lack of fresh water and scurvy took their toll.

The 16th century rivalry between Portugal and Spain sparked what was perhaps the bravest single voyage in the history of exploration, the first circumnavigation of the planet, commanded by Ferdinand Magellan.

A minor Portuguese noble born in 1480, Magellan fought in his country's bloody campaigns against its Muslim commercial rivals up and down the Indian coast, from Goa to the spice port of Calicut. In 1513, Magellan was wounded fighting the Moors in Morocco, an injury that left him lame for life. While he was in command of the rear area, jealous rivals falsely reported he had dishonored Portugal by selling cattle to the enemy. Although he was vindicated, the slur hurt Magellan worse than the physical wound. His honor sullied, he abandoned soldiering to devote himself to exploration and commerce. But when he asked King Manuel to back an expedition to the Spice Islands, the monarch rebuffed him. Magellan boldly replied he might seek support elsewhere. "Serve whom you will, Clubfoot," Manuel scornfully answered.

Magellan studied the latest findings of European geographers, some of whom speculated there might be a water route to the Indies south of Vespucci's American continent. He went to the Spanish court at Valladolid, renounced his allegiance to Portugal, and petitioned King Charles I for support on this new expedition.

Magellan's argument was persuasive. The Spice Islands—which he had visited during the India campaign—undoubtedly lay west of the Papal Line; even though Portugal had reached them first, the rich archipelago was therefore rightfully Spanish. If he sailed west around America and continued directly through the Indian

An allegorical view of Ferdinand Magellan in 1520 passing through the strait that bears his name. Guided by nymphs, but threatened by monsters, Magellan reached the Pacific 38 days after entering the strait. At left are the fires of Tierra del Fuego.

Ocean, no one could dispute Spain's claim. The Spanish king agreed. Magellan was named Captain General in charge of a five-ship fleet. Unfortunately, King Charles appointed several inexperienced noblemen to command the four other vessels.

Magellan's fleet left Spain on September 20, 1519. After three difficult months at sea, they finally anchored on the Brazilian coast in the bay of present-day Rio de Janeiro. The signs of scurvy had now appeared. Magellan rested his disgruntled crews, then ordered the fleet down the coast to search for the sea passage to Asia. By mid-January, they had reached a wide gulf—today's Rio de la Plata—and Magellan sailed upriver a few leagues. But the waterway was obviously not a saltwater channel to the Indian Ocean.

The fleet pushed south down the coast. Winter was approaching, the sailors and their Spanish captains were weary and restive, and the ships needed repair. Magellan decided to spend the cold season in the sheltered anchorage of Bahia San Julian, in the far south of present-day Argentina. The sailors traded with indigenous people who wore bulky hide boots, which resembled *patas* (paws). Magellan named the coast Patagonia.

At midnight on Easter 1520, Spanish captain Gaspar de Quesada led a mutiny. But Portuguese sailors loyal to Magellan helped him quell the revolt. Acting boldly, Magellan ordered Quesada executed.

Finally, in October 1520, Magellan took his diminished fleet down the endless coast once more, searching for the passage west. The *San Antonio* went ahead to scout a promising bay. But the ship's captain turned back for Spain. Undeterred, Magellan led his three remaining ships into a narrowing channel. As the channel widened and narrowed, harsh westerly gales swept down, churning off the headlands to strike the ships with sails aback. Some days they lost ground and were forced to anchor in exposed coves. As they skirted the southern coast, the sailors saw signal fires lighting the ridges. In his log, Magellan described this shore as Tierra del Fuego.

Thirty-eight days after entering the strait that would later bear his name, Magellan led his three ships to a protected anchorage beneath a headland he called Cape *Deseado* (Cape of Desires). Magellan's stern self-discipline faltered. He sobbed with joy, having proven there was a water route into this great Southern Ocean. Since the sea seemed calm after the violent whirlpools and downdrafts of the strait, he named the new ocean the Pacific.

Confident of quickly reaching the Moluccas by sailing northwest, Magellan set off across the vast Pacific, unaware of the distance that lay between the South American coast and the East Indies. The three small ships sailed for weeks, never sighting land. The salted seal and penguin they had prepared went bad in the equatorial heat. Scurvy took its toll.

In late January 1521, Magellan swept his chart table clear and threw the expensive vellum scrolls into the hot blue sea, cursing the cartographers who had arbitrarily placed the Spice Islands near his present position. "The Moluccas," he said in despair, "are not to be found at their appointed place."

Almost every day a sailor was buried at sea. Some men gnawed on boiled leather to ease their hunger. If the sailors had been able to navigate their vessels home, they would have mutinied. But they were lost, on the far side of a huge planet, and there was no recourse but to continue the voyage.

Then, on March 6, after more than 100 days at sea, with the vessels' provisions completely exhausted, they sighted a large island with green hills continuing well inland. Smoke from cook fires rose among the villages along the coral beaches. As the sun-bleached ships dropped anchor, outrigger canoes sped from shore carrying healthy and smiling people, eager to welcome the strangers to their land. Magellan had crossed the immense Pacific and reached the island of Guam, largest in the Marianas.

The islanders' generous hospitality saved Magellan's sailors from certain death, as coconuts, yams, and fresh greens quickly reversed their scurvy. But, lacking a European sense of private property, the natives filched whatever piece of metal or tool they found. The sailors named the islands the *Ladrones* (Thieves).

Ten days later, Magellan anchored in the present-day Philippines. His servant, Black Henry, who had followed his master since the India campaigns, spoke Malay, as did the local people. Magellan parlayed with the raja of Cebu, who—in an expedient quid pro quo—he converted to Christianity in exchange for a military alliance. Honorable soldier that he was, Magellan kept his word and joined the raja on a punitive expedition against a rival on the neighboring island of Mactan. Leading a band of 50 Europeans armed with pistols, firearms, and crossbows, Magellan encountered a much larger party of native warriors. As the sailors tried to retreat to their boats, Magellan was hit by poison arrows. He refused assistance and continued to wield his sword while his surviving men withdrew. The last they saw of their Captain General, he was lying face down, warriors hacking him with stone axes.

Pioneering Portuguese explorer Ferdinand Magellan sailed for Spain's Charles I, charting the sea route around South America to the Pacific and onward to the rich Spice Islands.

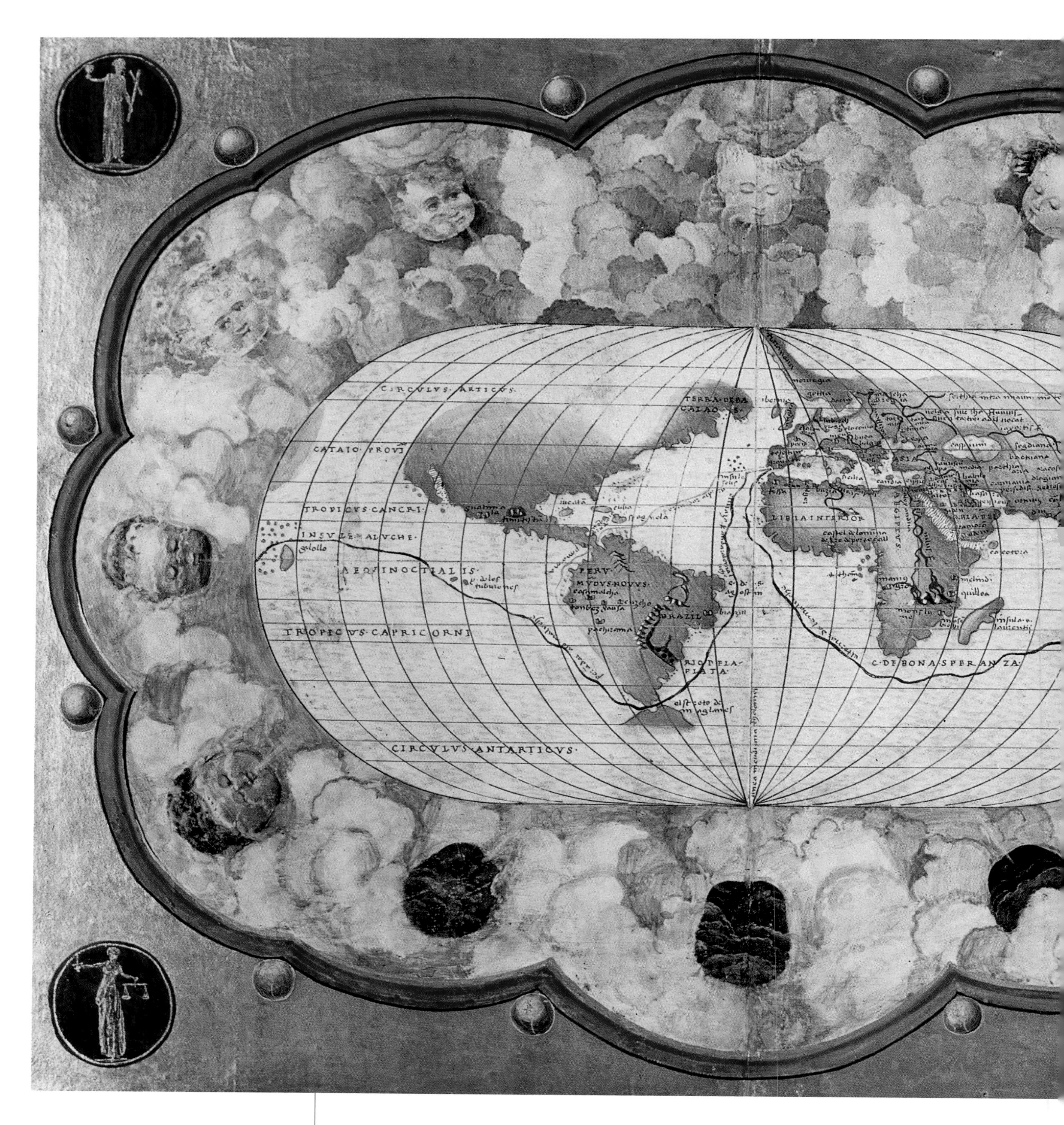

The expedition that had begun with such noble intent, now degenerated into piracy when the survivors elected Joao Lopes Carvalho, a Portuguese officer, as their commander. Finally, in November 1521, led by a new Captain General, a Spaniard named

Juan de Elcano, the two remaining ships reached the Moluccas where they successfully traded for a cargo of spices.

With a handful of survivors aboard the *Victoria*, Elcano left the East Indies for the Cape of Good Hope. He eventually reached Spain in September 1522, almost exactly three years after Magellan's expedition had departed. Although Magellan himself did not survive this first circumnavigation, when the account of his incredible voyage became public, geographers could at last make reliable estimates on the size of the planet, the southern extent of the American continent, and the true scope of the oceans.

During this first circumnavigation, Magellan lead a five-ship Spanish fleet. They sailed in 1519 seeking to round South America, cross the Pacific, and claim rich Asian possessions for King Charles I. Magellan battled mutiny, scurvy, starvation, and savage seas to reach his goal, but once in Asian waters, he was killed fighting islanders in the present-day Philippines. Only one ship completed the voyage, crossing the Indian Ocean and rounding Africa, but Magellan's triumphant exploration transcended his death.

In early April 1975, Columbia University's Bruce Heezen, a world-renowned marine geologist, asked me to stand in for him during an expedition to the Panama Basin on the Pacific side of the isthmus. He had to attend a petroleum-geology conference in Texas to receive a prestigious award. Good for him and good for me, I thought, ordering my plane tickets. The cruise would employ the Navy submersible *Turtle*, a clone of *Alvin*, and the chartered civilian tender *Maxine-D* to investigate seafloor features.

I would fly to the Colombian Pacific port of Buenaventura via Miami and Cali to join the ship. The first week of April on Cape Cod can be bleak, the dregs of mud season; Professor Heezen didn't have to ask me twice.

I didn't complain about lugging over a hundred pounds of awkwardly clunking gear for the ship and bulky boxes of 70 mm film for the submersible. But when I stood at the customs counter in the musty Cale airport terminal late on April 3, I began to have second thoughts.

The customs man, a mustachioed bureaucrat out of B-movie Central Casting, scowled at the film boxes and grunted as he tried unsuccessfully to heft one corner of the twine-wrapped carton with the gear.

"What do you have inside of there, Señor?"

"Spare parts," I said, with a weak smile.

"For an automobile?"

"For a friend," I managed.

He shook his head dubiously. "...for a friend."

I stole a look at my watch. The Air Zapata connecting flight to Buenaventura was scheduled to leave in 20 minutes. If I didn't make it, I'd be stuck in Cali all night.

Now he was trying to peel back the heavy foil wrap of the 70mm film.

"Please be careful," I said. "That's film. I'm a tourist"

He looked me up and down. "Where is your camera?"

"Here's my passport," I said, handing across the blue-gray rectangle. Inside, I'd folded three twenties, something I had heard about but never attempted. What if the customs man took the money, then called the cops for attempting to bribe him? I could imagine what the local jail was like.

A smile appeared beneath the mustache as he expertly palmed the money and stamped my customs form. "Where will you be a tourist?"

"In Buenaventura. I have to take the Air Zapata flight immediately."

Now he frowned. "Air Zapata, Señor? They have not flown airplanes for many years. No airplanes fly to Buenaventura."

A grinning porter pushed my mountain of baggage outside to the deserted sidewalk and harangued me in machine-gun Spanish. The only words I understood were "taxi" and "Buenaventura."

Half an hour later, I was dozing in the back of a big old rattling Chevrolet taxi. The night air of the Valle del Cauca was cool. The hillsides looked fertile in the moonlight. We drove past silent warehouses, pungent with coffee. I would drink some strong espresso in the morning; now I'd grab a little sleep.

The flashlight was hot in my face. In the glare I saw the rifle muzzle wavering about six inches from my forehead. Another weapon thrust into the rear seat from the opposite window. The cab driver had his hands raised meekly. What was this, a police roadblock?

"*Norte American*," I mumbled in pidgin Spanish. "Geologist."

The guy with the carbine in my face swept the light around the back seat, the beam lingering on my briefcase. I saw that all the men at the roadblock wore red armbands replete with the acronyms of their communist guerrilla group. On the flight down from Miami, I'd heard Colombian passengers referring to "*La Violencia*" as if it were some type of disease. Now I understood the meaning of the words.

"*Tu dinero*," the man at my window growled, thrusting the weapon closer. "*¡Ahora!*"

I gave him my wallet, my hand shaking badly. He took the cash, tossed the empty billfold back into the car, and waved us through. Fortunately, I had stashed enough local currency to pay the taxi driver inside my shoe back at the airport.

Now I was awake. The cool air of the upper Cauca valley gave way to rank tropical heat and humidity. Flying termites smacked the windshield. The night stank of rotting vegetation. Somehow, I managed to sleep again.

I awoke with flies on my face. The taxi was parked on the Buenaventura shrimp wharf in a smoky dawn of burning garbage. Around me, the walls of the 16th century port were crumbling, overgrown with creepers in places. The water of the Rio Dagua rose with a greasy tide.

The *Maxine-D* was not tied alongside the commercial pier where it was supposed to be. I reached down to retrieve the local currency I had wedged between my sock and shoe. It was gone. I must have lost it wrestling with the baggage at the airport.

"*Momentito*," I soothed the driver in what I hoped sounded like Spanish. He wouldn't go anywhere until I paid him.

Bruce Heezen's instructions had been to find the American consulate if I had any problems. The building was up the hillside. I started hiking.

First, I had to find my way out of the maze of dank old fortifications around the harbor. The limestone-block walls were pitted from the cannonballs that pirates and assorted rebels had fired over the centuries. As I negotiated the narrow lanes between between the double walls, whores hissed and grasped at me from the shadows. Once I was free of the walls, stevedores sipping their early morning coffee surrounded me, begging for the chance to lug my baggage from the taxi to the pier.

Soaked with sweat, my chest heaving, I reached the American consulate 15 minutes later. The metal gate was locked. A plaque announced business hours began at 8:30 a.m. By now I was surrounded by a pack of eager street urchins.

One of them pointed to the river below. "*¡Tu barco!*" he exclaimed. My ship, the *Maxine-D*, was coming up the river, rounding the green point of Cascajal Island. I turned back down the hillside and ran the gantlet of stevedores and whores once more. Now there were armed military guards at the gates of the walls, young kids in jungle fatigues with automatic weapons. I tried to look resolute as I mumbled something about "*el barco*."

Thankfully, the taxi was still waiting on the concrete shrimp wharf. I assured the driver as best I could that he would be paid, then made my way down the dock as the ship maneuvered alongside.

"I'm going to need 40 bucks to pay my taxi," I called, standing on the superstructure.

He waved that he understood. Everything would turn out okay after all, I thought.

Then a crewman handed a line to one of the street urchins on the dock. The boy snatched the sailor's watch and dove in the water. A dugout canoe full of whores slid up to the opposite side of the ship and the girls began to clamber aboard, assisted by a few eager sailors. The docking maneuver was going straight to hell.

A nervous young officer grabbed an M-16 rifle and fired into the air. This was not a good idea in a country wracked by years of violence. People on the wharf hit the ground. The military guards in the old walls leveled their weapons toward the ship. I felt terribly exposed out there on the dock. But the tense moment passed. Tempers cooled. The sailors completed their work.

I paid off the patient driver and found a bunk in a reasonably well air-conditioned berthing compartment. All I wanted now was eight hours of uninterrupted sleep.

But that proved impossible. I woke to feel a strange weight on the mattress and rolled to one side. The weight was still there. Opening my eyes, there was a shadow moving above me. I flicked on the light. A young Colombian whore, wearing the same gaudy clothing and makeup as the ones in the dugout, was perched at the end of the bed searching my empty wallet.

I reached out my hand and she meekly passed it to me with a shrug. "Sorry," I said. "Somebody beat you to it."

When we sailed next morning, the jungle closed down to the riverbanks, and there was a pervasive scent of mildew and decay. I stared at Buenaventura's crumbling stone walls.

V Ocean Highways

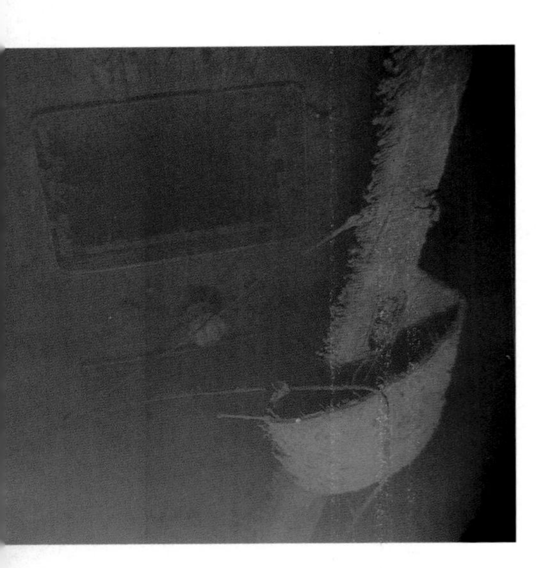

Preceding pages: In 1986, we explored the *Titanic*'s grave, which we discovered the previous year, using the submersible *Alvin* (upper left). Here *Alvin* rests on the ship's bridge, while the small ROV *Jason Jr.* (right) inspects the fallen mast.

The spectral image of *Titanic*'s cylindrical crow's nest from which lookout Fred Fleet had sounded the alarm, "Iceberg right ahead!" greeted us on the screen. The massive ship, steaming at 22 knots, could not avert disastrous collision.

Like a bug-eyed alien, *Jason Jr.* investigates stalactite rustcicles that have formed on *Titanic*'s port anchor over the decades since that tragic night in April 1912.

My fingers gripped the *Knorr's* bow rail as I stared toward the storm front smearing the Atlantic sunset. It was Saturday, August 31, 1985, the fifth week of the joint French-American quest to find *Titanic*. In five days, our ship charter would end and we'd have to leave the search site, 310 miles southeast of Newfoundland. And all we had discovered so far on the seafloor 12,000 feet below was flat gray mud and ripply sand, punctuated by the occasional sea slug.

As chief scientist, I had to face the probability of failure. My colleagues from the Institut Français de Recherches pour l'Exploitation des Mers, led by my old friend Jean-Louis Michel, had spent a fruitless month aboard the French vessel, *Le Suroit*, scouring the area with their ultra-sensitive side-scanning SAR sonar before joining my Woods Hole Oceanographic Institution party. For the past eight days aboard *Knorr*, we had searched unsuccessfully for *Titanic* with our sophisticated towed video sled, *Argo*. The immense, tragic ship had to lie nearby, eluding us, just as it had previous expeditions. Making my way aft along the starboard side, I passed the jutting crane with *Argo's* thin tow cable slicing straight down as *Knorr* crept across the surface. The sled was gliding just above the seafloor, its sensitive video eyes completing yet another search track. In the blue control van, the mission's nerve center, the people on watch hunched at their stations, their normally optimistic mood tempered by the prevailing low morale. Soon that long cable would be reeled up for the last time, *Argo's* jaunty white Fiberglas tail fin would emerge from the sea, and the ship would head west toward New England.

And the press would be waiting, demanding to know why the world's best technology had produced such dismal results. For me, the questions would have a more personal sting. I had recklessly boasted that finding *Titanic* would be a "relatively easy task" once I obtained funding for high-tech gear and assembled the right team. Well, there were no better search tools than the French SAR—which produced detailed sonar shadowgraphs resembling black-and-white photographs—or *Argo*, which could visually scan long swaths of ocean bottom with its unblinking Silicone Intensified Target video cameras. And our people represented the cream of undersea exploration.

I suddenly wondered if the optimistic media hyperbole on both sides of the Atlantic had actually jinxed the expedition. But I shook my head at the illogical thought. Still, at low points like this, it was easy to believe the great ship's sad legacy would defeat anyone audacious enough to search for her grave.

Titanic's legend certainly rested on audacity. The ship's near mythical size, almost 900 feet from sharp-angled bow to gracefully contoured stern, and displacement of over 45,000 tons, made it and its sister ship, *Olympic*, the largest vessels ever launched at that time. And White Star Line captain Edward J. Smith was an audacious seaman. On April 14, 1912, during the ship's maiden voyage from Europe to New York, Smith had disregarded a wireless message from a Cunard liner that "bergs, growlers, and field ice" straddled his planned course near the Grand Banks. Even though other captains chose to drift during the clear, moonless night, Captain Smith disdainfully steamed ahead at 22 knots.

His boldness was tragically rewarded when an iceberg gouged the starboard side below the waterline, and the double-bottomed hull of the "unsinkable" liner flooded. When *Titanic* tilted nearly vertical and slipped, creaking and rumbling,

We launch the *Sonar Acoustique Remorqué* (SAR) near the end of the French phase of the 1985 *Titanic* expedition. Although unsuccessful, the SAR's effort greatly narrowed the search area for the subsequent hunt with the American towed video sled *Argo*. We later discovered the SAR had passed within 1,000 meters of *Titanic*'s debris field before the stowed video/sonar sled *Argo* took over the search.

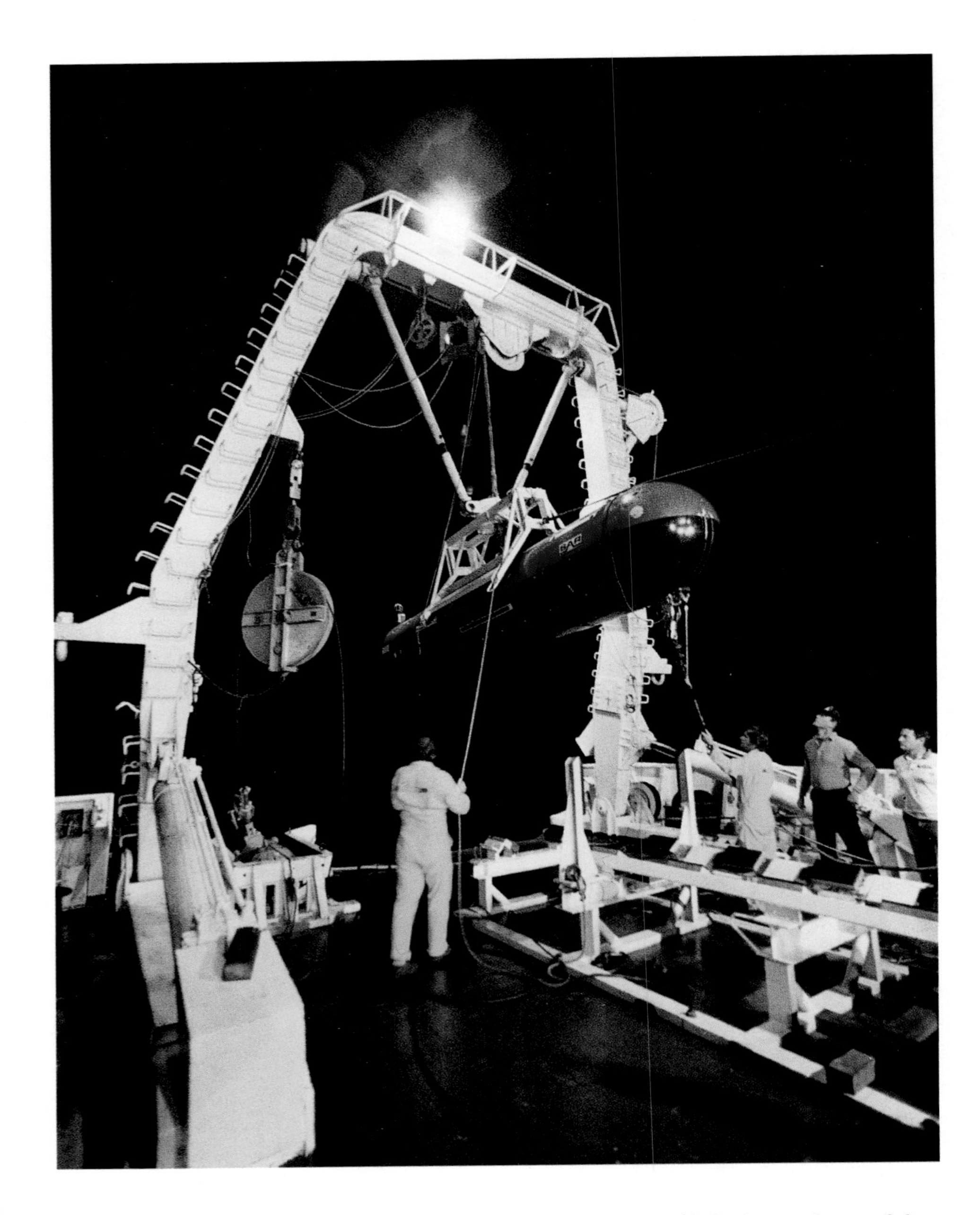

beneath the icy Atlantic at 2:20 a.m., Smith must have tasted the bitter dregs of that brashness. But his imprudence was not unique. Convinced their liner was invincible, the White Star Line had not installed enough lifeboats for all the 2,200 passengers and crew. More than 1,500 died.

Because the ship had been the epitome of the Edwardian era's technological prowess and devotion to conspicuous luxury, the maiden voyage had drawn the elite of the British aristocracy and their American counterparts. A one-way passage in a handsomely appointed First Class suite would have cost a workingman a year's wages. But the sinking taught a harsh lesson: Despite immense wealth and prestige, an Astor or a Guggenheim could drown just as easily as an anonymous Polish immigrant from steerage.

A ship sank that night and the boundless faith in the benign power of technology that had grown throughout the 19th century was also mortally wounded. Four years later, in the mechanical hell of the Western Front, that optimism was finally extinguished.

Argo, the vehicle we used to discover the wreck of the *Titanic* in 1985, ran the search lines at wide intervals—"mowing the lawn." *Argo*'s unblinking video and sonar eyes relentlessly hunted for debris around the clock.

Following pages: At 2:20 a.m. on April 14, 1912, the White Star Line's pride *Titanic*, sailing on her maiden voyage, struck an iceberg and sank, drowning more than 1,500.

Now, savoring my own pending failure, I considered the events that had set me on this unlikely voyage.

As a teenager in Southern California reading the gripping account, A Night to Remember, I was enthralled by the image of *Titanic*, her Grand Staircase, domed skylight, and crystal chandeliers eternally preserved, a mummy of a gilded age, beneath the impenetrable ocean. At Woods Hole in the 1970s I came to realize that *Titanic* might not, in fact, lie beyond human reach. Alvin was due for a new titanium pressure sphere that would permit it to dive past 12,000 feet, the depth in which the great liner sank. But the little submersible was a poor search vessel, so I envisioned a sophisticated towed video camera-and-sonar sled that would allow us to "fly" search runs across the ocean bottom while seated comfortably and safely in a control van aboard ship.

Argo was born from this concept. By 1984, I'd secured funding from my old

Audacious and visionary British ship designer Isambard Kingdom Brunel launched his globe-spanning steamship *Great Eastern* in 1845, intending the ship for nonstop voyages around Africa to Australia. By far the largest ship yet built, much of the hull was taken by coal bunkers.

in the 1800s and extended rule over almost the entire Indonesian Archipelago, the little Netherlands was one of Europe's richest nations.

From the establishment of British colonies in North America until the 19th century, many free European settlers arrived aboard small privately owned merchant sailing vessels known as packets. My own ancestor, Colonel Thomas Bal-

The 692-foot *Great Eastern* mounted two sidewheels and a massive propeller. But smaller ships, which fit through the new Suez Canal, robbed *Great Eastern* of her anticipated trade. Brunel's "great babe," however, did lay the first Atlantic telegraph cable in 1866.

lard, landed at the James River port of Williamsburg in the Virginia Colony in 1635 on just such a ship. The crossing was often rough, generally uncomfortable, and frequently deadly when epidemic struck the dank, poorly ventilated below-decks. The American Revolution, soon followed by England's long wars with France, dampened transatlantic immigration.

In this period, however, two American inventors, John Fitch and Robert Fulton, developed technology that would not only transform *continued on page 150*

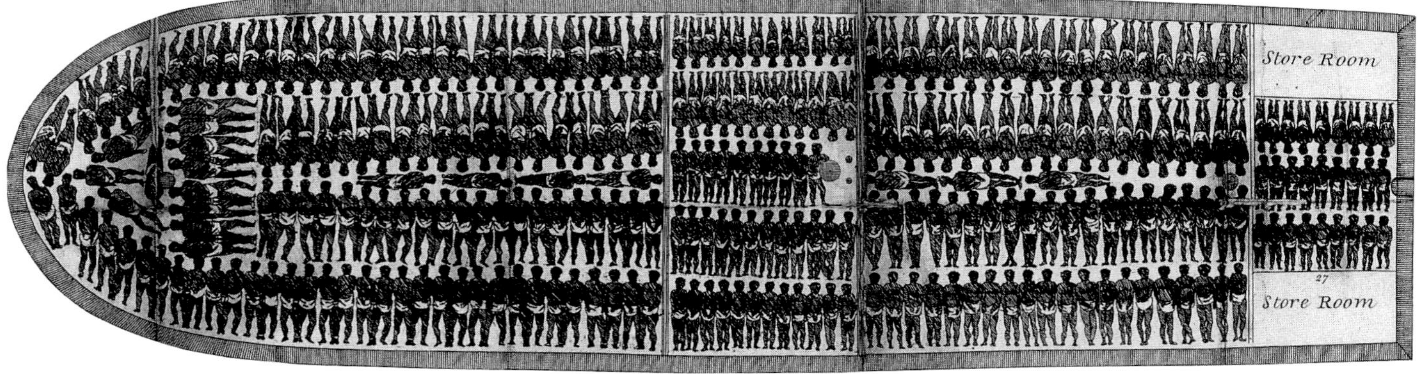
Store Room
Store Room

Immigrants in Chains: Slaves and Convicts

One of the cruelest chapters in history unfolded on the Atlantic during the 300 years between the 16th and 19th centuries: Spanish, Portuguese, British, French, and Dutch slavers carried uncounted millions of African captives from their home continent to the New World on the dreaded Middle Passage, a voyage synonymous with misery and death.

The institution of slavery predated Western civilization, but the multistage commerce known as the triangular slave trade took the practice to a new nadir. The triangular trade was named for the three-legs of the slaver's voyage. The transport of African slaves across the Atlantic was started by Spanish and Portuguese colonists in the 1500s. It wasn't until vast plantations of sugar, tobacco, and cotton were later created on the Caribbean islands and in the British North American colonies, however, that a ceaseless demand for slaves arose.

Typically, a slave ship would depart from a European port laden with trade goods, often including barrels of rum distilled from Caribbean molasses and bolts of cheap print cloth woven from New World cotton. Next the slaver would anchor at one of many "factories" dotting the Slave Coast on the Gulf of Guinea. Local tribes traded captives they had seized during forays to the interior for the slaver's trade goods. This pitiful human cargo was herded aboard and jammed into low, stifling holds, the men chained together at the ankle to discourage mutiny, the women and children less restrained. All were crammed onto tiers of platforms filling the holds like shelves. There was no room to sit upright or even turn. Ventilation was minimal, sanitation nonexistent.

The hold plan of the Liverpool slaver *Brook* in 1781 allots each adult man slave a narrow stifling space, jammed in shoulder-to-shoulder, head-to-toe with his fellow captives; women and children had even tighter quarters. Designed to transport 500 slaves, the ship sometimes carried over 600; its master anticipating inevitable "wastage" of the valuable human cargo. There was little ventilation and no sanitation. Death from disease and abuse was common. To prevent rebellion, newly captured men slaves were shackled. Still, many rose against their tormentors, choosing death rather than continued suffering.

On the third leg of the triangular trade, slave captains filled their holds with barrels of molasses or bales of cotton, which would be processed into rum or cloth in Europe. The brutal process came full circle when the slavers sailed once again for Africa.

On the actual passage, dysentery took its toll as the ships, many carrying up to 600 slaves, made the slow voyage from West Africa to the Caribbean or Colonial American ports. Experienced slaver captains always estimated an average of around 20 percent "wastage" when they provisioned their vessels with minimal rations of millet, yams, or corn meal. Every day, the bodies of the dead were unceremoniously dumped overboard, sharks trailed the slavers. Most ships tried to preserve their valuable cargo by keeping slaves on deck during daylight and "dancing" (exercising) the morose captives with the encouragement of the lash.

Mutinies aboard slave ships were a constant hazard to both crew and captives: There are more than 50 well-documented accounts of shipboard rebellions between the 17th and 19th centuries. Often the slaves were massacred with grapeshot from swivel guns; sometimes the slaves bludgeoned especially cruel crewmen and threw them into the sea. The most famous slave mutiny occurred on *Amistad* in the 1830s. Slaves took the Spanish schooner on the Cuban coast, and the crew eventually sailed the ship to American waters. Former President John Quincy Adams successfully argued the slaves' case before the United States Supreme Court, which denied a suit to return them to their Spanish masters. The survivors of the *Amistad* became freemen.

But African slaves were not the only chained captives crossing the Atlantic. Britain's

YORK
Portsmouth

Transportation Act of 1718 established the practice of emptying prisons by banishing convicts to the distant American colonies. Prisoners were often given the dubious choice of involuntary servitude overseas or the gallows. Over 50,000 chose transportation. As capital offenses ranged from poaching to simple theft, there was no shortage of convicts put to work clearing forests and draining marshland in the colonies. Then came the American Revolution, and the British Crown needed another destination for their convicts.

Establishing a new penal colony near Botany Bay in Australia became the practical, if harsh, alternative. While convicts crossing the Atlantic had faced a voyage of perhaps two months, those transported to Australia could easily be aboard ship four times as long. And even before the 11 ships of the First Fleet were assembled in Portsmouth in May 1787, many convicts had languished aboard filthy unheated hulks for months or even years. But Royal Navy Captain Arthur Phillip maintained good order over his convoy during the 250-day voyage to the far side of the planet. Only a handful of convicts and crew died.

Soon, however, the British government turned the transportation fleet over to private contractors, some former slavers. Between 1787 and 1868, 825 ships left Great Britain carrying over 160,000 convicts to Australia. The convicts' plight was often little better than that of Africans on the Middle Passage. Chained below deck, frequently soaked with icy seawater, they were infested with lice and tormented by boils. Sanitation was primitive at best. Corrupt captains cut back on rations to sell their remaining provisions in Australia at immense profit. Dissent, especially among Irish politicals, was met with barbarity. One captain, Thomas Dennott, took sadistic pleasure in flogging convicts to death during his ship's 1796 passage from Cork.

Convicts board a 19th century British prison hulk, a disused warship. Some languished aboard for years before transport to Australia. After the Crown contracted slavers to transport prisoners, barbarity ensued.

In 1993, we explored the wreck of R.M.S. *Lusitania*, which the German Imperial Navy submarine *U-20* sank off the Irish coast in May 1915. ROV pilot Martin Bowen (center) maneuvers *Jason*, while Bob Elder (far left) works the robot's computers. Artist Ken Marschall and I use a model to search for recognizable objects on the wreck.

the Atlantic passenger trade, but also lead to the fast ocean liners and cargo ships that knit the continents into a single global economy a hundred years later.

I learned in school that Robert Fulton invented the steamboat. In fact, that honor rightfully goes to John Fitch, a veteran of the Revolutionary War who settled near the Delaware River. Hoping to adapt a version of Scottish inventor James Watt's steam engine to a boat, Fitch tried unsuccessfully to obtain funding from the Continental Congress in 1785. So Fitch turned to private backing to build a 45-foot steam paddle wheeler, which he successfully demonstrated on the Delaware. The Philadelphia press lauded Fitch's vessel, which deposited "passengers with great regularity at either extreme of its course." Unable to raise enough money to invest properly in his invention, Fitch went to France, hoping to interest the science-conscious

Republican government in his project. Again, he failed. He died in 1798, his invention largely forgotten. None of the books I ever read as a boy mentioned John Fitch.

Robert Fulton was born on a Pennsylvania farm when Fitch was 22. Fulton studied painting in England, but his fertile imagination leaned more toward technology than the arts. An intellectual explorer far ahead of his time, he dabbled in marine engineering, but could not convince European navies to purchase his submarine *Nautilus*. However, his work in France put Fulton in contact with Robert Livingston, American minister to Napoleon's court. They formed a partnership to develop a small steam-powered paddle wheeler similar to Fitch's first boat. The French were impressed, but did not invest.

In New York, Livingston had used political connections to secure a steamship monopoly, provided Fulton could build a vessel that attained four miles an hour, then an amazing speed for a river craft. I can picture the mixed fear and skepticism on the faces of the spectators lining the Manhattan dock on August 17, 1807, when Fulton's famous *Clermont*, a 150-foot steam paddle-wheeler, puffed and snorted up the Hudson toward Albany. The ship actually departed without exploding and arrived in just 32 hours, almost four times as fast as sailing boats. Livingston maintained his patent. Steamboats were proven practical. And sailing ships' long reign, which had lasted from at least the mythical days of Homeric heroes and ships such as *Tanit* and *Elissa*, had heard the first notes of its death knell.

In May 1819, a 300-ton sail packet, *Savannah*, which had been hastily converted to mount two belching steam engines and paddle wheels, left its homeport in Georgia bound for Liverpool. The ship was basically a sail vessel with auxiliary steam engines fueled by tons of Georgia pine logs. When the wind was fair on the 24-day crossing, the captain used sail. But when it dropped, he furled sail, got up steam, and churned ahead. Near Ireland, a rescue boat sailed out, convinced the dark smoke roiling from the ship's stack meant uncontrolled fire.

The first ship to make the Atlantic crossing completely under steam power was the *Sirius*, a small two-masted side-wheeler, whose paddles were lauded in the press as "gracefully shaped and painted black all over." In May 1838, the vessel carried 40 passengers from Great Britain to New York. Encountering fierce head winds, the crew became restive, and her skipper, a Royal Navy lieutenant named Richard Roberts, brandished a big revolver to quash the incipient revolt. He drove hard to beat a larger rival steamship, *Great Western*. But *Sirius*'s 400 tons of coal were exhausted within miles of New York harbor. Rather than raise sail, Roberts brusquely ordered the masts and yardarms to be sawed apart and fed into the boiler to keep up steam past the Sandy Hook Light. His ship had crossed the Atlantic in a record 17 days.

New Yorkers cheer as R.M.S. *Lusitania* sails for Liverpool on May 1, 1915, despite German warnings the ship would enter U-boat-patrolled waters near England. *Lusitania* was torpedoed six days later.

***Following pages:* The *U-20*'s torpedo struck below the starboard waterline, triggering a massive secondary explosion. *Lusitania*'s momentum plowed her forward and the ship sank in just 18 minutes, drowning 1,195 of the 1,955 people on board, including 123 neutral Americans.**

During our close-in inspection of *Lusitania*'s hull 78 years after the sinking, one of our goals was to discover if the ship had carried contraband munitions as some had long claimed. Although the wreck lay on its starboard side, concealing the site of the torpedo impact, we found coal strewn on the bottom and concluded that *Lusitania* sank when coal dust exploded in an empty bunker.

Canadian-born Samuel Cunard joined with British partners in 1839 to establish the British and North American Royal Mail Steam Packet Company. The firm, soon known as the Cunard Line, provided the first regular Atlantic steamship service. During this period, Swedish engineer and inventor John Ericsson overcame metallurgy problems to perfect a practical screw propeller that eventually replaced paddle sidewheels. The age of the ocean steam liner was born.

As steam technology advanced, new side-lever engines began to shake apart wooden-planked vessels. Yet the unprecedented potential speed of these engines was too great a prize to forgo. Visionary British ship designer Isambard Kingdom Brunel tackled this technical problem with characteristic panache. His fabulous ship the *Great Britain* launched in 1845, perhaps the first true ocean liner. Its hull was riveted iron plate, and contained watertight bulkheads and a double bottom. It was the first ocean vessel driven by screw propeller, not paddle wheels. But the ship was not a commercial success. It ran aground off Ireland in 1846, and although its 180 passengers were rescued and the ship was later refloated, its owners went bankrupt.

However, Brunel remained a bold thinker. His next ship, the *Great Eastern*, was superlative in both concept and design. Brunel envisioned a vessel that could carry up to 4,000 passengers to India and on to Australia. This monumental voyage would require immense coal bunkers in the hull, the reason for the unprecedented length of 692 feet (230 meters) and displacement of 32,160 tons. Brunel's ship, which he lovingly called his "great babe," mounted two big side wheels and a gigantic propeller. It also carried sail on six tall masts. Technical problems plagued the vessel, and it entered the more mundane Atlantic passenger service rather than its planned globe-spanning route. But its distinctive and perhaps greatest accomplishment was laying the first Atlantic telegraph cable in 1866, which linked North America with Europe in a nearly instantaneous communication web.

Although steam liners would retain their masts and rigging for much of the century, their interiors evolved radically from earlier sailing ships. As I discovered studying these vessels, the most expensive cabins were aft, in the driest, most stable part of the hull. But, beginning with the White Star Line's *Oceanic* in 1871, luxury accommodation moved forward, away from the noise of the engines and rumbling propeller shafts. I saw the designers' logic: Given the full width of the hull to work with, they built spacious cabins, luxurious lounges, and dining rooms for their First

Class passengers. This left unused accommodation room aft, in "steerage," where ships' machinery was located.

As I also learned studying maritime history for the exploration of the liners *Titanic*, *Britanic*, and *Lusitania*, British liners would evolve to reign supreme on the Atlantic. Cunard and its rival White Star Line produced increasingly fast and luxurious ships. Strong steel hull plate replaced iron. Multiple screw propellers now drove the vessels. First Class dining rooms boasted two-story atria domed with skylights. Acres of gleaming mahogany and polished mirrors graced the public rooms. Now America's wealthy class on their Grand Tours to the Continent comprised many of the passengers.

A wartime British enlistment poster draws on the indignation over *Lusitania*'s sinking to incite anti-German fervor. Similar outrage almost drew the United States into World War I in 1915.

But Germany challenged British dominance of this extravagant commerce. It's indicative of the period that, after visiting the new White Star liner *Teutonic* in 1889 and noting the impressive amenities, which included a barber shop with an electric hair dryer, Kaiser Wilhelm II commented, "We must have some of these." Within a decade, the German liner *Kaiser Wilhelm der Grosse* captured the coveted Blue Riband for the fastest Atlantic crossing from the British. But the ship was not only fast, it was the epitome of unrestrained lavishness. The high ceilings of her public rooms were gilded with ornate carvings, the plaster walls presented allegorical bas-relief. Glowing beaux-arts stained glass celebrated the ship's pompous grandiosity. It and its German sisters soon became the most fashionable vessels on the North Atlantic run.

To counter this competition, Cunard's chairman, Lord Inverclyde, convinced the British government in 1902 to loan the line an unprecedented 2.6 million pounds for two grand luxury liners that would regain the Blue Riband from Britain's German rivals. The *Mauretania* and the *Lusitania* would be 750 feet long and displace over 31,000 tons, making them the biggest vessels since Brunel's exotic monster the *Great Eastern*.

Although the ships combined the expected mix of first class opulence and spartan steerage accommodation, it was their propulsion that was truly exceptional. To achieve optimal speed, Cunard gambled by replacing proven piston engines with still-experimental steam turbines. Designed by a gentleman engineer named Charles Parsons, turbine marine engines used steam at maximum efficiency. But the risk was that these turbines, which delivered an amazing 68,000 horsepower, might use too much coal to make the vessels profitable. This was pushing technological exploration to its limit.

However, the engines proved efficient. And, during *Lusitania*'s speed trials in July 1907, she cruised steadily at 26 knots. That year the new liner took back the Blue Riband from Germany, crossing the Atlantic from Ireland to New York in under five days. *Mauretania* soon beat her sister ship's crossing by 24 minutes. For the next seven years, the two ships were friendly rivals, establishing such a record for speed and reliability that they earned the unofficial title "the Atlantic ferry." Unfortunately, the rivalry among the European powers did not remain limited such peaceful pursuits. In 1914, the World War I erupted and the Atlantic became a battlefield.

While *Titanic* was sunk indirectly by human foible, *Lusitania* was destroyed by a direct, merciless act. On May 7, 1915, ten months into World War I, the liner was cruising among fog banks off the southern coast of Ireland on a return voyage from New York. Captain William Turner was confident his vessel's incredible speed protected her from German U-boats that had been reported off the Irish port of Queenstown. Turner adjusted course and increased speed as the ship broke free of the fog. His new heading took him directly toward the U-20 commanded by Kapitanleutnant Walther Schwieger of the Imperial German Navy.

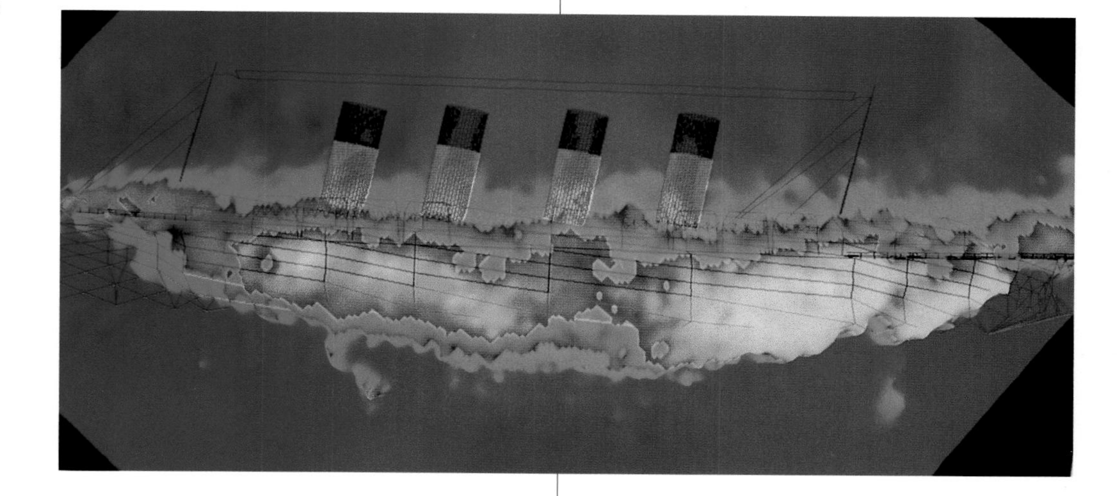

With *Jason*'s high-frequency sonar, we created a computerized image of the wreck and overlaid it on a scale diagram of the actual ship. We found *Lusitania*'s flattened hull had broken between the third and fourth funnels, a weak spot similar to one on *Titanic*, which suggested a design flaw.

The submarine fired her single remaining torpedo, which struck *Lusitania* below the starboard waterline near the bridge. The initial explosion triggered an immense secondary blast, ripping a long gash in the hull. Still plowing ahead, the mighty ship listed hard over and sank within 18 minutes in 295 feet of water. The death toll was horrendous, 1,195, including 123 Americans. Germany claimed the ship was a legitimate target because Cunard had disregarded warnings that the waters around Great Britain were a war zone. And some saw the secondary explosion as proof that *Lusitania* had been secretly carrying tons of contraband munitions from neutral America. Others created a wild conspiracy theory in which Captain Turner had purposely led his ship into the U-boat trap so that her sinking would inflame American opinion against Germany.

The tragedy did become a cause celebre, and eventually contributed to the United States' decision to enter the war beside the Allies in 1917. There was no evidence, however, that Turner had participated in the tragedy. But why had the great liner sunk so quickly? The mystery lingered for much of the 20th century.

I explored the *Lusitania*'s wreck in 1993, using the ROVs *Jason* and *Homer*, as well as the small research sub, *Delta*, deployed from the *Northern Horizon*. One of the goals of our Woods Hole expedition was trying to determine the cause of the terrible secondary explosion that had mortally wounded the liner that morning in May 1915.

Our initial ROV video survey revealed the liner lay on its starboard side, its hull plates collapsed like tin shacks after a gale. From our perspective the hull's position was a disappointment: the starboard side partially hidden beneath a jumble of deck wreckage and collapsed superstructure. Worse, cloying spider webs of ripped fishnet made close-up inspection hazardous to both ROVs and the minisubmarine *Delta*.

But we persevered to conduct a complete sonar and video survey of *Lusitania*, even though three members of the expedition—artist Ken Marschall, historian Eric Sauder, and pilot Chris Ijames—were briefly trapped when *Delta*'s rudder and propeller were snarled in fishnet. The wreck was too deep for scuba divers to descend to attempt a rescue. Chris, who had logged over 2,000 dives, struggled to free the

Using the ROVs *Jason* and *Homer*, we carefully explore *Lusitania*'s shattered, fishnet-shrouded hull. The submersible *Delta* works further aft. Later, *Delta*, with artist Ken Marschall, historian Eric Sauder, and pilot Chris Ijames on board, became entangled in fishnet. Unable to escape, Ijames had to jettison his rudder and propeller section.

submersible. Finally, he had to resort to jettisoning the rudder-propeller section, an emergency expedient wisely incorporated in the vessel's design.

We rushed to the ship's rail, scanning the slate gray Irish Sea anxiously. Then the *Delta* shot to the surface like a lemon yellow cork, surrounded by a gleaming cloud of bubbles.

"Thank God they made it," I said. "We've got to be a little more careful, gang."

Now we pursued our exploration with even greater caution. *Lusitania*'s starboard hull was bulged in a rupture below the waterline, evidence of the secondary explosion's violence. But the section of the ship containing the ship's magazine, a fireproof hold where the small cargo of officially authorized munitions had been stored, was intact. Close inspection with *Delta* and the ROVs, however, revealed the seafloor near the gashed hull was littered with chunks of coal. The U-20's torpedo had penetrated a nearly empty coal bunker in the ship's side, churning up tons of volatile dust that must have ignited, transforming the compartment into a gigantic bomb.

I am satisfied that our careful exploration probably solved one of the most enduring mysteries of the 20th century. *Lusitania* had not been transporting tons of explosive munitions, as some have claimed, only small quantities of properly manifested war materiel such as three-inch artillery fuses that had been stored in the ship's magazine, which, we verified, had remained intact.

World War I marked a hiatus in the golden age of transatlantic travel. But after that bloody conflict, the Roaring Twenties saw a resurgence of glittering luxury liners, plying routes between America, Great Britain, and the Continent. The French now challenged British preeminence with ships such as the *Ile de France* and the *Normandie*. But America's mood of postwar isolationism, which had sparked anti-immigration legislation and reduced the flow of steerage passengers to a trickle, left the great ships traveling with a mere handful of passengers.

It was middle-class American tourists, with vacationing college students in the vanguard, who became the salvation of the transatlantic liners. Travel agents convinced steamship companies to offer their spartan accommodations as "tourist class." Soon, the lines refitted their ships with a hierarchy of cabins below First Class, ranging from small four-bunkers for families to floating dorm rooms for frat boys in raccoon coats.

This configuration continued after World War II. "Going to Europe" aboard a liner like the *United States* or the *Queen Mary* became almost commonplace for many Americans. Then long-range commercial jets appeared. Ocean liners, which had survived challenges from U-boats and stifled immigration, were finally superseded by the Boeing 707 and the DC-8. By the mid-1960s, regular Atlantic liner service had all but ended.

But the glamour of shipboard life is certainly not extinct. Today cruise ships have become the most popular form of tourism. On these glitzy floating hotels, people can still book luxury suites rivaling the *Titanic*'s opulence, though now all the passengers share the dining rooms, show lounges, and casinos, and polo shirts and shorts have replaced silk brocade and white tie. But passengers still experience the thrill of standing on scrubbed deck planks as the ship rumbles through the dark ocean, its phosphorescent wake mirroring the starry sky.

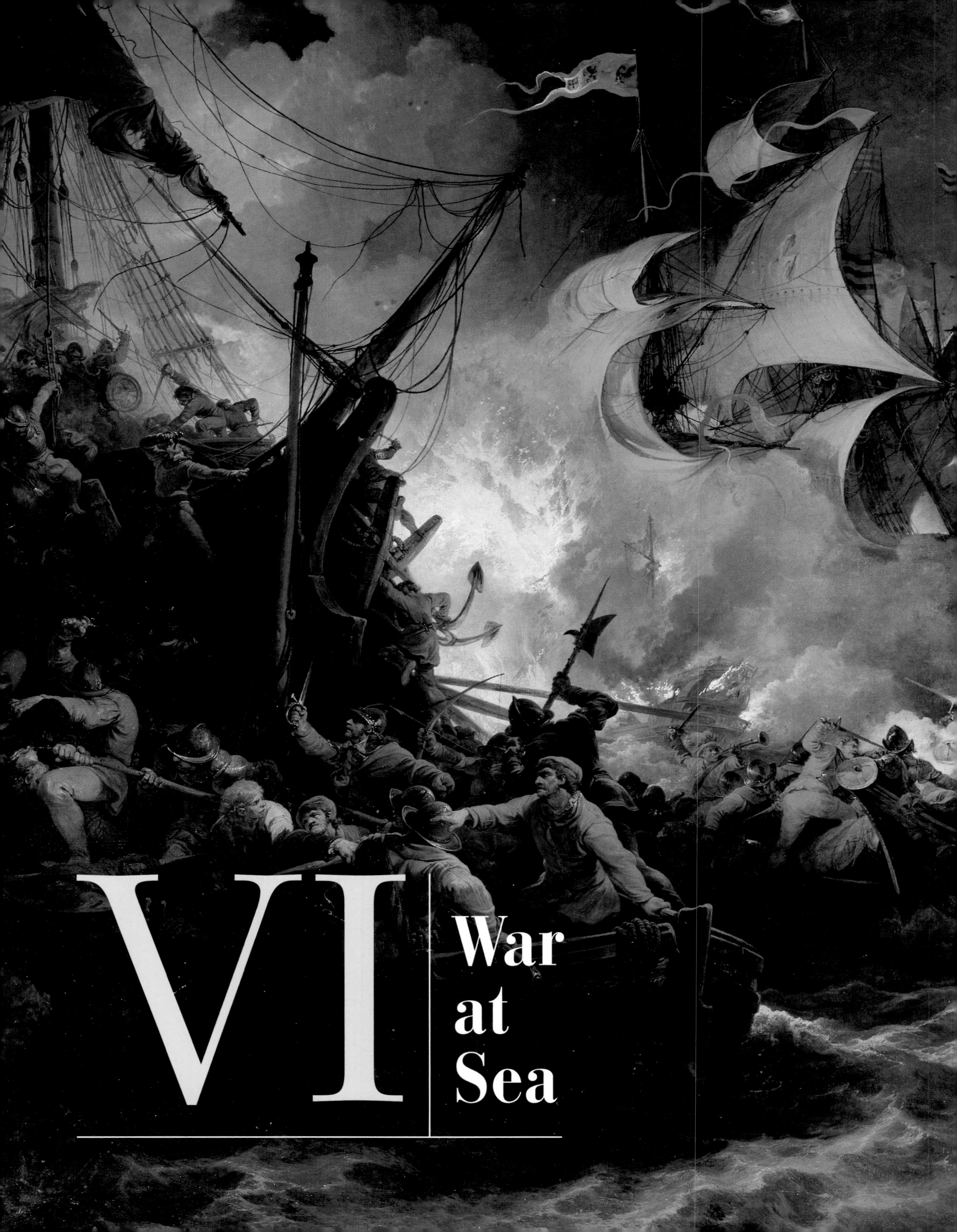

VI
War at Sea

Preceding pages: English fireships plow into the Spanish Armada anchored on the French coast in August 1588. Many Spanish sailors panicked, cutting anchor lines and scattering in the night.

A single 15-inch shell from the Nazi battleship *Bismarck* sinks the British battle cruiser *Hood* (background), while its consort the *Prince of Wales* dodges wreckage. Over 1,400 sailors died in seconds. Days later, *Prince of Wales* helped sink *Bismarck*.

It was near midnight on June 5, 1989, but sleep would be hard tonight. Wedged into a Naugahyde booth in the mess hall of the *Star Hercules*, playing a round of Trivial Pursuit, I was haunted by the parallels between the German battleship *Bismarck* lying somewhere on the Atlantic seafloor 15,700 feet below, which had so far eluded me during two expeditions, and the *Titanic*, which had almost evaded us four years earlier. Both ships had been proclaimed as technical marvels, one the epitome of Edwardian Pax Britannica's opulence, the other the most powerful weapon in Nazi Germany's *Kriegsmarine*. They were both commanded by blindly overconfident officers. Both had been touted as unsinkable, yet both had sunk on their maiden voyage with terrible death tolls.

I had hunted *Bismarck* without success the previous summer, and now had been scouring the slopes of a rugged volcanic seamount rising from the Porcupine Abyssal Plain southwest of Ireland with our camera sled *Argo* for eight fruitless days and nights. Our search area overlapped the precise sinking position of the warship, as recorded by the navigators of the British Royal Navy ships that had finally sunk the enemy vessel on May 27, 1941. Still, a giant battleship, 820 feet long, displacing almost 45,000 tons, remained invisible to *Argo*'s sonar and video cameras.

I knew from British reports that the *Bismarck* had sunk with her hull intact. Given its thick armor, which had withstood shelling for hours during the climax of the most savage Atlantic surface battle of World War II, I was certain it had reached the bottom in one piece. Why couldn't my team at least find a debris trail?

The *Star Hercules* vibrated as we turned slowly to starboard, beginning another search track, our unblinking little vehicle at the end of its almost three-mile cable, flying 90 feet above the seamount's shoulder.

On the night of May 19, 1941, the powerful new battleship *Bismarck* rendezvoused with its heavy cruiser consort *Prinz Eugen* in the Baltic and

headed toward the North Sea. Only recently completed, *Bismarck*'s rakish but thickly armored hull evoked speed and power. It could maneuver at 30 knots, outpacing the older battleships of the British Home Fleet that might attempt blocking the German task force's passage into the open Atlantic where the ships had orders to prey on vulnerable supply convoys from North America.

Bismarck was ideally suited for this assignment. Its eight 15-inch rifled naval cannon mounted in four turrets could fire one-ton shells almost 20 miles. And the battleship was equipped with the latest radar and stereoscopic fire directors that could adjust devastating salvos with cruel accuracy. *Prinz Eugen* and *Bismarck* also carried Arado floatplanes to scout for convoys at long range so that the warships could dash up unseen and fire from beneath the horizon.

Should *Bismarck* encounter British cruisers or battleships, the German vessel's potentially vulnerable boilers, turbines, and ammunition magazines were encased

In June 1989, we searched for the Nazi warship *Bismarck*, towing *Argo* from the research vessel *Star Hercules*. After endless hours hunting with video and sonar, we finally encountered debris on the side of a seamount, 16,000 feet below. We soon found the massive wreck itself.

in a thickly armored interior fortress. The decks and turrets were clad in the strongest tempered armor plate that German foundries could produce; the plates could withstand direct hits from all but the largest British naval cannon. Bearing the proud name of Imperial Germany's "Iron" chancellor, Otto von Bismarck, the battleship was the paradigm of German naval architects, "an unsinkable gun platform."

The threat *Bismarck* and her consort posed to the vital convoy lifeline was a dagger poised above Britain's heart. In May 1941, the British Empire was a belligerent still fighting the Axis. German forces controlled Europe from Arctic Norway to Egypt, where Field Marshal Erwin Rommel's Afrika Korps was threatening the Suez Canal. *Luftwaffe* planes ranged far into the Atlantic, bombing hapless merchant vessels and coordinating U-boat wolf pack attacks on convoys. If the German submarines and air force, augmented by heavy warships, destroyed the North Atlantic convoy system, Great Britain would starve. Prime Minister Winston Churchill's government would be forced to sue for peace. Fascism would reign triumphant in Europe.

That was the audacious goal the Kriegsmarine had given Admiral Gunther Lutjens, commanding the German task force. He planned to sail northwest from Norway, above Iceland, then slip down the Denmark Strait near eastern Greenland and move further into the Atlantic to rendezvous with a tanker and replenish his vessels' oil. An RAF reconnaissance Spitfire photographed the *Bismarck* and *Prinz Eugen* in Norway's narrow Grimstadfjord. Now it was certain *Bismarck* had "broken out," but the Royal Navy did not know the enemy's intentions. And even with a fleet as numerous as Great Britain's, the Atlantic was too vast to saturate with ships.

Luck stayed with Lutjens. Low clouds and fog shrouded the Norwegian Sea all the way to Iceland, precluding successful British air reconnaissance. The German ships steamed south close to the Greenland pack ice, threading a narrow gantlet west of the British minefield stretching seaward from Iceland. But the weather cleared on the night of May 23; the *Bismarck* and *Prinz Eugen* were spotted by patrolling Royal Navy cruisers, *Norfolk* and *Suffolk*, which then shadowed them with radar.

The battleship *Prince of Wales* and the heavy battle cruiser *Hood* sped northwest to intercept Lutjens's task force. In theory at least, the opponents were perfectly matched. In reality, the edge in long-range gunnery went to the Germans. The opening salvos were fired as the ships closed at dawn light on May 24. Before any British fire struck home, one of *Bismarck*'s 15-inch shells smashed through *Hood*'s deck and exploded deep inside, igniting tons of munitions in a magazine. The ship erupted like a volcano, breaking in half and sinking within seconds. Among the crew of 1,419, only three survived. *Bismarck*'s guns had killed the equivalent of an army regiment in seconds. The damaged *Prince of Wales* fled.

But the German ships, which had also been hit in the exchange, could do nothing to shake the two British cruisers dogging their trail, just beyond range of the big guns. Obsolete Swordfish torpedo planes from the aircraft carrier *Victorious* attacked *Bismarck* in the rosy midnight-sun twilight on May 24, inflicting only minor damage. In the confusion of the battle, Lutjens ordered *Prinz Eugen* south to pursue convoy raiding, then doubled back on his own track to shake the shadowing British cruisers. Reluctantly, Captain Robert Ellis of the *Suffolk* radioed the message, "Have lost contact with enemy."

Part of the *Bismarck*'s debris was this lonely sea boot, all that remained of a German sailor killed in the sinking. Nearby were haunting clusters of boots from men who had drowned, their life jackets tethered together.

Free of pursuit, the German admiral headed southeast toward Occupied France, and the protective umbrella of long-range Luftwaffe bombers. *Bismarck* was low on fuel from battle-damaged tanks and had suffered a flooded boiler room. She could only steam at reduced speed. Still, Lutjens was confident he could safely reach dry dock in Saint-Nazaire for repairs. He unwisely broke radio silence, and *Bismarck*'s messages were tracked by the British.

Early on May 26, an RAF Catalina flying boat managed to penetrate *Bismarck*'s flak barrage and transmit an accurate report on the ship's speed and heading. Almost 12 hours later, lumbering Swordfish biplanes from the carrier *Ark Royal* closed on the battleship just above the ragged storm swell from two directions. Two torpedoes exploded harmlessly on the thick armor of *Bismarck*'s hull. But one blasted into the vulnerable twin rudders, jamming them hard to starboard. The German ship turned in uncontrollable circles to port while engineers struggled to make repairs, and Captain Ernst Lindemann unsuccessfully attempted to maneuver using variable thrust on the propellers.

Inexorably, *Bismarck* swung north toward Ireland, away from the Luftwaffe's protective umbrella. The pursuing Royal Navy tightened the noose on the German warship after dawn on May 27. Rolling through a harsh gale, 300 miles southwest of Ireland, the battleships *Rodney* and *King George V* opened fire as the range closed rapidly. Unable to maneuver, *Bismarck* returned fire ineffectively. It was struck by hundreds of 16-inch and smaller caliber shells over the next two hours before all its turrets stopped firing and the ship went dead in the water. Smoke poured from the blazing superstructure as German sailors grouped around life rafts or simply leapt into the chill water. Captain Lindemann ordered scuttling charges fired to sink the ship. The heavy cruiser *Dorsetshire* delivered the coup de grace torpedo attack, then stood by with the destroyer *Maori* to rescue enemy survivors.

The crews of the British vessels watched with grim satisfaction as *Bismarck* settled steadily at the stern, then rolled over to port, its flat red bottom paint showing briefly before slipping beneath the surging swell. It was 10:39 a.m. on May 27, 1941. The cruiser and destroyer began pulling shocked and wounded Germans from the cold, oily water. But only 115 survivors had been lifted aboard when an urgent U-boat warning was flashed to the British task force. As several hundred horror-stricken German sailors watched, their life jackets tied together in long lines, the British ships zigzagged away at high speed. The men floating in the water died of exposure before the end of the day. In all, the 110 survivors already rescued were the only men left alive among the crew of 2,206 that had left the Baltic nine days before. The *Hood* had been avenged.

The Battle of the Atlantic continued between convoy escorts and U-boats for the next three years. But the Kriegsmarine never again sent large surface ships to threaten the vulnerable North Atlantic convoy lifeline.

The seamen of the *Dorsetshire* exalt after delivering the *Bismarck* its final blows. Following the torpedo attack on the *Bismarck*, the heavy cruiser and its crew stood by to retrieve enemy survivors.

I felt a hand on my shoulder and looked up at my old Woods Hole geologist buddy, Al Uchupi. "Bob," he said with his Bronx accent, "we've encountered some debris I think you should have a look at."

I ran flat out for the control van, scattering potato chips in my wake. Although Al had spoken in a noncommittal tone, I knew he'd chosen the word "debris" with care. In the van, *Argo*'s watch crew gazed at a clump of small, smudgy black objects drifting diagonally across the monitor. I'd seen enough debris trails by now to recognize this material as man-made. I nodded to the watch navigator, Cathy Offinger, who was printing neatly in the logbook, "debris." It was 23:52 hours on June 5, 1989. Scanning the VCR playback monitor of the debris that had brought Al to the mess hall, I recognized scattered chunks of narrow pipe and twisted shards of steel.

"If it's *Bismarck*, we've got two possibilities," I muttered, as much reasoning with myself as addressing the crew. "Either this stuff was shot off during the battle, or it's part of the sinking debris field."

Then Mel Lee, the sonar operator, said, "We're not picking up any contacts on the side-scan."

That confirmed my suspicion: *Argo* had drifted over a scattering of light battle-damage trash blown from *Bismarck*'s superstructure sometime in the ship's final two hours. The objects on the screen probably did not present the predictable winnowed trail that would lead us to the main wreck. An hour later, the scene abruptly changed. The gray snowdrift mud gave way to a violently disturbed pattern of mixed rock and sediment, as if some unbelievably giant hand had pounded the slope with a monster sledgehammer. Were we looking at the lip of *Bismarck*'s impact crater? Then we were back on smooth, featureless mud, without encountering any more debris.

In frustration, we turned *Star Hercules* in a slow arc to the west, finally encountering more scattered light debris. Doubling back to the east, *Argo* followed a trail of small twisted metal objects into more torn-up mud. The tension in the van was palpable: The battleship *had* to be there. But all we found as we cleared the lip of the crater was a gentle mud slope devoid of debris.

Watching the near featureless video monitor, a somber thought took hold. Maybe *Bismarck*'s armored hull had impacted with such force that the entire wreck was buried in the bottom sediment. Perhaps we'd already "discovered" all the wreckage we ever would.

Then Al Uchupi recognized the significance of the chewed-up landscape we'd been searching for the last several hours: It was a muddy avalanche that had rolled down from the upper seamount. As he explained his theory, I realized that *Bismarck*'s 45,000 tons, smashing into the slope at high speed provided more than enough energy for this phenomenon. Somewhere along the length of the landslide, I hoped, the ship itself would lie. Now we just had to use the axis of the avalanche to search in a narrow ladder of *Argo* sweeps until we located the hull. That was the theory at least.

A partially severed swastika air-recognition panel painted on the *Bismarck*'s aft deck, as photographed by *Argo* during our 1989 exploration, supplies the first sight of this hated image reminding us of the Nazis' powerful tyranny.

Following pages: **Quadruple antiaircraft gun mounts aboard the heavy cruiser U.S.S. *Quincy*, sunk in Iron Bottom Sound off Guadalcanal during the 1942 Battle of Savo Island. Exploring the sunken American and Japanese ships after the battle, the shattered hulls bore witness to the savagery of the combat.**

The endless sweeps produced tantalizing evidence, but not the wreck itself as *Argo* slowly climbed the flank of the undersea volcano. The video monitor revealed a ripped sheet of welded steel and a section of ladder. Then the chilling image of a solitary boot lying in the churned sediment drifted across the screen. *Argo* climbed higher up the avalanche path. No major debris, only occasional clumps of minor wreckage. I was exhausted, having not slept for over 30 hours, and was still gripped by almost paranoid certainty that the main hull lay beneath the mud.

"The sediment's not thick enough for that, Bob," Al Uchupi reasoned. He argued the ship would have careened down the seamount with the landslide, remaining above the dislodged bottom material. We reversed course and worked our way toward the lower slope.

About halfway down the mudslide, we encountered a big cluster of boots—all that remained of the survivors bobbing together, their lifejackets attached by lanyards. I felt a chill. The humanity of these objects seized me. How long had they waited in the icy water for rescue that never came?

Now the hours became a numbing smear. I dozed between cans of Coke. Each time I rubbed my face awake, the stubble was thicker. Still, no *Bismarck*. Jack Maurer's watch was back on duty. I had missed dinner again.

"That looks like a skid mark," Al exclaimed, pointing to the monitor.

"Oh, wow!" Kirk Gustafson shouted from the flyer station. "Look at that!"

The circular image was crisp and unmistakable, a ring of large gear teeth, the tracking mechanism of a 15-inch gun turrets. Hundreds of Royal Navy witnesses had seen *Bismarck* roll over on sinking. The big "barbette" turrets would have broken loose and fallen separately from the hull, the heaviest chaff in the debris trail. I measured the diameter of the image on the screen with a plastic ruler and managed a plodding calculation despite my fatigue. We had located an object eight meters (26 feet) in diameter. "That's a main turret, lying upside down," I said with certainty.

But another frustrating day passed as *Argo* found several more pieces of wreckage, including a chunk of superstructure lined with portholes, but no sign of an intact hull. I was lying groggily in my cabin on the morning of June 8, watching the repeater monitor as I debated between a much-needed shower or breakfast. Suddenly the endless gray mud gave way to the stark image of two naval cannon jutting from an angular turret, and the dark vertical line on the screen's left margin was not just another steep gully. It was the edge of *Bismarck*'s armored hull. "We've got it!" I yelled, running toward the van in my stockinged feet.

Six hours later, I sat beside *Argo* flyer Billy Yunck gripping a plastic model of *Bismarck*. *Argo*'s sensitive SIT cameras flew less than 20 meters above the battleship's shell-scarred decks. *Bismarck* lay upright and nearly intact in a hollow on the lower slopes of the seamount 15,700 feet below. Its sharply raked prow pointed southwest, away from its elusive sanctuary in Saint Nazaire.

The four big turrets were gone, leaving holes like the sockets of missing molars in a fossil jawbone. But the smaller-caliber gun turrets were still in place. The ship looked remarkably ready for combat. We found the worst shell damage on the portside, amidships, where one of *Rodney*'s 16-inch shells had pierced the armor. Scores of German sailors must have died in that explosion. Gazing at the edges of the shell-ripped hull, I could almost hear the voices screaming in fear and anger, their words lost in the exploding chaos.

But that shell gap was the only major visible breech in the heavily armored hull. The ship's compartments must have already been flooded before it reached implosion depth. Here was evidence supporting the German survivors' account that Captain Lindemann had ordered *Bismarck* scuttled before the crew abandoned it.

As the downward-looking zoom lens moved aft across the still-intact teak decks, strange angular markings emerged. I'd seen similar discolored patches near the bow, undoubtedly paint over wood. Now I recognized the image: a wide swastika air-recognition insignia meant to identify the battleship for the Luftwaffe. That hated symbol of Nazi tyranny had been expunged from Europe, but rested here on the deep ocean floor, mute testimony to the defeat of Hitler's aggression. I looked around the control room at the faces of my colleagues who stared somberly at the swastika. We certainly hadn't hunted for *Bismarck* to glorify Nazi Germany. Rather, our exploration had been a quest for truth, an attempt to add another page to the long history of war at sea.

Warfare and seafaring have been linked since Paleolithic people started lugging stone axes aboard their dugout canoes. By the Homeric period, ship had become synonymous with warship. For centuries these vessels carried raiding parties to enemy shores, as in the siege of Troy. However, direct warfare among armed vessels did not become common until the relatively recent historical past. In the last three millennia, sea battles have often been decisive turning points that shaped world civilization.

I received an unusual perspective on such warfare when National Geographic Television producer Christine Weber, filmmaker Peter Schnall, and I were invited to inspect a replica of a classical Greek trireme on the Aegean island of Poros in August 1987.

According to the history I'd read, triremes were the fastest, most maneuverable, and most destructive warships of the ancient world. The lean, light vessel had evolved from earlier war galleys that traced their lineage back to the age of mythical heroes such as Jason. As nautical technology advanced, one bank of oarsmen gave way to two—the bireme—and two banks became three. Triremes were about 120 feet long, but only 18 feet wide. Constructed with the thin tongue-and-groove planking of the period, the ships weighed less than 40 tons.

Triremes of the Persian and Greek fleets clash near Salamis in 480 B.C. This etching portrays the warships carrying sail, when in fact they relied solely on oar power during battle. The Greek victory changed Western history by expelling Persia from the region.

In the fifth century B.C., triremes had achieved unparalleled speed through the unique configuration of their rowers. Between 150 and 170 oarsmen were seated in three tiers on each side. In the larger triremes, there were 27 rowers in the bottom and middle tiers and 31 on the top, which was extended from the ship's hull by an outrigger. Two square sails could also be rigged from foldable masts, but oars were the sole propulsion in battle. The rowers were staggered both vertically and horizontally so that their 14-foot oars all bit the water with maximum leverage yet did not touch.

According to the Trireme Trust, an English group that had designed and built the replica *Olympias* with the cooperation of the Greek Navy, the vessels could reach speeds of almost ten knots in brief sprints, making them the fastest machines of the era. The official speed trial down at the Greek Navy base at Poros was the reason Chris, Peter, and I were waiting in the jostling crowd of backpacking college students and Greek island villagers assembled on the Pireás hydrofoil dock this bright, windless morning.

As the rumbling, futuristic vessel rose on its foils, diesel plumes snorted from its stacks. I stood at an open hatch savoring the artificial breeze. The bug-like craft sped southwest across the glassy water of the Saronic Gulf. Behind us, Athens-Pireás disappeared in the brown dome of smog. Off to the starboard, I saw the green-and-tan slopes of Salamis Island rise from the shimmering evaporation haze.

After quickly reaching Poros, the rest of the hot morning passed with the inevitable slow mechanics of naval protocol and film production. By afternoon, we were ready to film the replica's official speed tests across a measured course outside the harbor. The crew were volunteers, young men and women from Oxford rowing clubs.

I watched with mounting excitement as they maneuvered the big, dark-planked ship into position. Then the oars lowered in unison and began to stroke. In the Aegean sunlight the trireme was so beautiful; I forgot how deadly a weapon it

The replica trireme *Olympias*, crewed by British volunteers, undergoes speed trials near Poros in 1987. While nearing 10 knots in spurts, the vessel could not maintain this speed, but the designers proved that the ancient trireme was a formidable warship.

had been. Its stern rose in a graceful curve like a swan's neck. The bow was brightly painted and displayed a pair of glaring eyes.

Jutting from the bow beneath the waterline was the trireme's principal weapon: a bronze-clad wooden ram. Triremes carried parties of up to 50 well-armed *epibatai* fighting men. Their assignment was to grapple, board, and capture enemy vessels. Ships like this were the *Bismarck*s of their time. That afternoon, the replica trireme almost reached the ten-knot barrier, but the crew was so exhausted by their effort that I doubted they could have fought very effectively afterward. And combat was the purpose of the warship.

As the hydrofoil carried us back to Athens on a scorching afternoon several days later, I saw the hills of Salamis grow distinct against the taller mountains of the Greek mainland to the north. For hours during the speed trials, we had watched the replica trireme churning back and forth, its wet oars flashing in sunlight. Now I almost expected to see more triremes, the ghosts of the epic conflict that had unfolded on these sparkling blue Aegean waters in 480 B.C., gliding toward us out of the dusk.

On the hot day of September 29 that year, the combined fleets of Athens and her Greek allies met the numerically superior forces of the Persian emperor Xerxes in the narrow channel between the island of Salamis and the mainland near Athens. At stake in the battle may have been no less than the future of the Western world.

In 480 B.C., the Persian emperor Xerxes invaded Greece with a slow-moving army that numbered several hundred thousand men and depended on seaborne supply lines. Cargo vessels carrying provisions were sluggish round-ship freighters. But Xerxes's fleet also had over 800 triremes to use as offensive weapons once the Persians had secured their supply-convoy routes. About two-thirds of them were crewed by Egyptians and Phoenicians. Many of the rest came from the Greek coastal cities of Asia Minor, which Persia had earlier subdued.

With news of Xerxes' pending invasion, the Athenian military leader Themistocles convinced his countrymen to invest a rich new lode of silver from the Laurium mines in a rushed shipbuilding campaign. That was perhaps the wisest investment ever made. Athens hoped to add 200 triremes to the proposed allied Greek fleet, which would include ships from Sparta, Corinth, and smaller cities and islands. In all, the Greek allies wanted almost 400 triremes manned with trained crews before the Persians reached their shores.

Propelled by banks of rowers, triremes slammed into their opponents' hulls beneath the waterline with their bronze-clad ram, the trireme's principal weapon. Marine infantry on board augmented the attack with javelins, spears, and arrows. Later, Greek fire (flaming naphtha) joined the arsenal.

As Athens prepared for the inevitable Persian onslaught, the city's precarious situation became clear. The new triremes had been hastily built with green timber because enough seasoned planks were not available. These hulls were heavy and slow. Worse, the newly hired crews were as green as the planks. They'd had no time to learn the intricate fleet tactics of the *diekplous* breakthrough or the encirclement meant to smash apart or trap an enemy formation. So the Athenian fleet had to rely on boarding rather than the complex maneuvers of the ram attack.

In the spring of 480, Xerxes' armies moved ponderously down the coasts of Thrace and Macedonia with the fleet following close offshore. The Spartans under their warrior-king Leonidas led the hoplite force of 5,000 in the heroic but futile defense of the mountain pass at Thermopylae. Although the Spartans were annihilated, the allied Greek fleet nearby at Artemisium slowed the advancing Persian ships. The Greeks must have recognized their weakness in ships and crew experience. If they were to win against the Persian triremes, the Greeks would have to fight them in narrow waters that restricted maneuver.

With the news of the Persian army nearing Athens, the city was convulsed with panic, which was worsened by a dire prophesy from the Oracle in Delphi. But Themistocles sent another envoy to Delphi. This time the Oracle's message was more reassuring: "Though all else shall be taken, Zeus, the all-seeing, grants that the wooden wall only shall not fall." Themistocles chose to interpret the "wooden wall" as the phalanx of triremes the city had so recently acquired.

But no miracle would save Athens itself. After a general evacuation, the victorious Persians marched in, burnt the ancient wooden temples on the Acropolis, and sacked the city. The Greek fleet was beached in coves on the nearby island of Salamis, its sailors watching helplessly as the eastern horizon glared red.

While the Athenian sailors wailed in anguish, Themistocles talked strategy with his allies. Several argued for a retreat down the isthmus toward Corinth. Themistocles reminded them that the enemy fleet could land troops behind any defensive line as long as Persian ships controlled the Aegean. The allied Greeks reluctantly agreed to accept his leadership.

VROOM 1617

artillery could fire solid iron balls weighing 50 or 60 pounds, but was only effective at short range. The weight of these guns, as well as their terrible recoil, also required that Spanish galleons be built extremely heavy, making them ungainly sailors. These galleons also carried large contingents of troops for defense and boarding parties. The Spanish saw their warships as floating fortresses, with the surrounding sea as the protective moat.

Spain's principal rival, Elizabethan Protestant England, explored new forms of war at sea. King Henry VIII introduced cannon that fired through individual gun ports cut in the hull. This allowed the English to place their heaviest cannon low, close to a vessel's center of gravity, which reduced aiming problems caused by roll. English tactics favored the lethal "raking" of an enemy's decks while passing perpendicular to his bow or stern.

These opposing tactics would be tested in one of the epic battles of Western history: the clash between Elizabeth's fleet and the Spanish Armada in 1588. With the Reformation, Europe was split between Catholics and Protestants, and Spain was the strongest, richest Catholic power. Queen Elizabeth's England had no empire and was poor by Spanish standards. Yet Francis Drake's bold strokes against the Spanish Main revealed that the English were determined to wrest their share of riches from the New World.

This was intolerable to King Philip II, as was the English support for Protestant Dutch rebels fighting the Duke of Parma in Holland, a Spanish province. Philip planned a strategy that would solve both problems: Dispatch a gigantic fleet to rendezvous with the Duke of Parma's army in Flanders and invade England with a combined force of over 30,000 men.

In May 1588, the fleet of Spanish and allied ships set sail, commanded by the Duke of Medina-Sidonia. He had marshalled over 130 galleons, carracks, and transports with a total armament of 2,400 cannon. His flagship, *San Martin*, displaced almost 1,500 tons and carried 500 men. There were 50 of these heavy galleons in the fleet, augmented by scores of smaller warships. The Armada also transported 8,000 sailors and 19,000 infantry, making it the most powerful naval force the world had ever seen.

Commanded by nobleman Charles Howard, whose vice admiral was Sir Francis Drake, the English put to sea hoping to intercept the Armada in the Bay of Biscay. But gales scattered both fleets. Damaged and short of supplies, Howard and Drake's ships limped back to Plymouth to refit. Legend has it that the two English admirals were playing a game of bowls on Plymouth Hoe on July 29, 1588, when word reached them that the Armada's vast crescent formation had been sighted off the Lizard in nearby Cornwall. Somehow, the Spanish had reassembled after the storm and reached the English Channel en route to Flanders quicker than anyone could have reasonably expected. Medina-Sidonia had caught the British fleet in the most vulnerable of possible circumstances. The southerly wind was behind the Spanish. The British ships were desperately short of gunpowder, shot, food, and water, and the strong tide was flowing directly into the harbor's narrow entrance, augmenting the headwind the English would have to overcome to escape the potential death trap.

But, according to the patriotic myth, neither English admiral panicked. Lifting his tankard of ale, Drake reputedly turned to Howard and said, "We've still time to finish the game and beat the Spaniards after." Plymouth harbor was soon a caldron of intense, but deliberate, activity. To overcome the adverse wind and tide, Howard ordered the captains of the 64 warships to "warp" out of the harbor. Each ship's longboat dropped an anchor at the extent of its cable. The crew manned the capstan, pulling the vessel against the tidal flow. Then the process was repeated until the ship had sea room to hoist sail and begin clawing to windward in zigzag tacks along the coast.

The English ships were more than equal to this demanding task. These "race-built" galleons differed markedly from their heavy Spanish counterparts. English galleons rode low in the water. Gone were the towering stern and fo'c'sle. Their hulls were leaner, with deeper keels. Their sails had less belly and could be sheeted flatter for better upwind sailing. These ships were less heavily armed than the enemy's lumbering galleons, but could maneuver with greater speed and agility.

Arriving off Plymouth, the Spanish sailors saw the glare of alarm beacons marching along the headlands. Francis Drake would have attacked, seizing advantage of the wind and tide, despite the harbor's fortifications. But Medina-Sidonia's strategic goal was linking up with the Duke of Parma's invasion force in Flanders and escorting his barges to England.

At sunrise on July 31, the Spanish looked to windward and saw small English formations—including Howard in his flagship *Ark Royal* and Drake aboard *Revenge*—darting across the freshening wind, which had shifted into the west. At each end of the Armada's crescent formation, like the sharpened tip of a scythe, big galleons held station, daring any British vessel to come within range of their 50-pounders.

On the English side, the pride in having extracted their ships from Plymouth gave way to dismay at the sight of the densely packed Armada, which stretched across the eastern horizon for six miles. Because there had never been such a large fleet, Howard and Drake had no tactical principles to follow in their attack. They decided to engage both tips of the crescent simultaneously to prevent the heavy galleons from providing mutual support.

After thunderous, but largely ineffective cannonades, Howard ordered his forces back to shadow the Spanish fleet. The Armada proceeded east up the Channel in bright sunshine. Over the next week, the English trailed the enemy, peppering them with long-range fire. Still the Armada plodded on toward Flanders. The wind dropped and the pursuing British fleet became desperate for gunpowder and cannonballs, dispatching schooners and fishing smacks to coastal forts to beseech the commanders for resupply.

On the dark, squally afternoon of Saturday, August 6, 1588, the Armada anchored in the shallows off the French port of Calais. Medina-Sidonia received news that Parma's barges were not yet ready. This meant the Armada must either remain at anchor in these dangerous shallows, move farther east into even worse shoal water, or initiate action against the British fleet to gain control of the Channel. Morale was slipping among the Spanish. The fast, skillful maneuvers of the English ships—and their lethally accurate cannon fire—had spread anxiety through the Armada.

It was under these conditions that the British loosed their surprise weapon on the night of Sunday, August 7. The English commanders took note of the steady westerly wind blowing onshore and that the tide would flow from the British anchorage toward the Armada near midnight. They decided to launch fireships against the Spanish. When the tide turned at 11:00 p.m., experienced Channel sailors steered eight fireships, loaded with wood and pitch, directly toward the

Sir Francis Drake, co-commander of the British fleet that defeated the Spanish Armada, saved England from invasion and altered the course of history. Drake's courage helped turn the battle against a superior force.

there following the American landings in August meant to dislodge the Japanese from this farthest southeast outpost of their empire. Many nights, the heavy Japanese warships of the "Tokyo Express" steamed up the channel to shell the Americans dug in around Guadalcanal's Henderson Field, the strategic airstrip the U.S. Marines had wrested from Japanese control.

When an especially strong Japanese task force centered on the battleships *Hiei* and *Kirishima* left Truk, guarded by a cruiser and a dense destroyer screen, the intent was to pound the American airbase so badly that there would be no planes capable of opposing the large troop landings timed to coincide with the attack. But an American B-17 bomber spotted the enemy task force en route to Guadalcanal.

For two nights savage, close-range naval battles raged up and down Iron Bottom Sound, with Japanese ships making especially effective use of their deadly Long Lance torpedoes. The first night of the engagement, November 12-13, American cruisers and destroyers were battered as they darted through the darkness near the Japanese battleships, raking their tall superstructures with armor-piercing shells.

As the ships' primitive radar antennas were blasted off by shellfire, their crews had to rely on searchlights and arching star shells, which gave the black tropical water an eerie glare laced with the wobbling orange tracers of cannon shells. By the end of the first night, several American destroyers and cruisers had been sunk. But the Japanese task force had also been wounded. The battleship *Hiei* had been damaged in its steering and engine room, almost like the *Bismarck*. It was steaming slowly in circles as morning came, easy prey to the "Cactus Air Force" from Henderson Field. After hours of bombing, the huge ship sank west of Savo Island.

That night, the surviving battleship, *Kirishima*, bravely led the Tokyo Express back into Iron Bottom Sound. But the U.S. Pacific Fleet had rushed in reinforcements, the battleships *Washington* and *South Dakota*, escorted by fresh cruisers and destroyers. Maneuvering in the narrow waters between Savo and Guadalcanal, the American force kept the Japanese ships from shelling the shore; *South Dakota* was badly mauled by *Kirishima*'s heavy guns. The *Washington*, however, was able to concentrate its 16-inch gun batteries on *Kirishima*, holing it below the waterline,

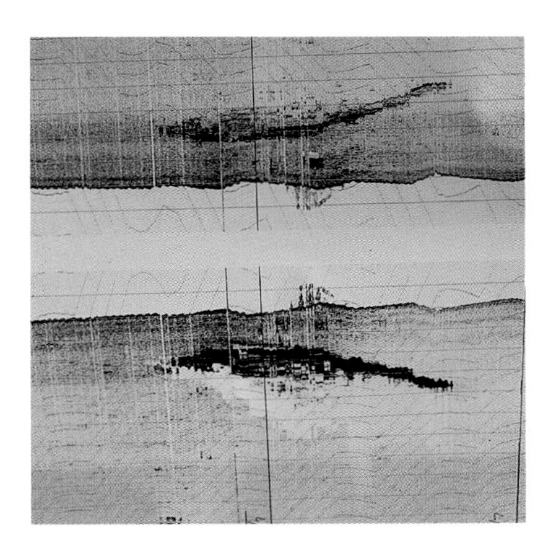

In November 1942, the American light cruiser *Atlanta* (left) engaged a superior enemy force off Guadalcanal. The fray became a close-quarters brawl and *Atlanta* was hit by both enemy and friendly fire. It was later scuttled.

To the north of the *Atlanta*, we found the American heavy cruiser *Northampton* lying upright in Iron Bottom Sound, as seen in this sonar image (above).

jamming the rudder, and knocking out two of its 14-inch gun turrets. The Japanese ship was mortally wounded. At about 3:30 a.m. on November 15, 1943, west of Savo Island, *Kirishima*'s towering superstructure toppled to starboard as its surviving crew plunged into the black, oily water. The huge ship, from which Emperor Hirohito had reviewed the Imperial Fleet before it sailed to the South Pacific, rolled over and sank. The last major Japanese navy threat to Guadalcanal sank with it.

After those savage engagements, American battleships were mainly used as floating gun platforms to soften up enemy beaches during the long island-hopping campaign on the road to Tokyo. But in those terrible, historic night engagements off Guadalcanal, more than 50 ships, large and small, had been sunk; thousands of men had been killed or wounded.

The Navy's *Sea Cliff*, a clone of *Alvin*, is launched from *Laney Chouest* off Guadalcanal during our 1992 expedition. We explored the largest deep-water collection of sunken warships, testimony to the savagery of the sea battles that raged there.

Almost 50 years after *Washington*'s 16-inch guns had sunk *Kirishima*, I was crouched in the clammy steel pressure sphere of the submersible *Sea Cliff*, a Navy clone of *Alvin*. We were on the floor of Iron Bottom Sound in 2,900 feet, creeping up on the huge sonar target of the sunken Japanese battleship *Kirishima*. But I was completely unprepared for what I saw outside the narrow viewport. Instead of our floodlights revealing the normal image of a shell-ripped but recognizable warship's hull and turrets, I saw a seemingly endless rusty, sponge-flecked expanse of curved steel.

Kirishima had turned belly up on sinking. Its towering armored superstructure acting like a keel, it had failed to right herself on the final plunge to the bottom. I was gazing at a mound of steel, not a once-proud and devastating weapon. It's as if the huge ship has been extinguished, I thought, not just sunk. *Kirishima* had crumpled into a hump-backed ridge of armor plate, scarcely recognizable as the powerful warship that had once been the scourge of the Pacific. As our submersible inched cautiously closer to the hull, I searched for familiar shapes on the wreck.

"My God," I whispered, "look at that."

We were above one of the battleship's four huge triple-bladed bronze propellers. In the floodlights, the upright propellers stood on their bearing braces like nightmare windmills. They remained in excellent condition, testimony to the skill of the Japanese foundrymen. Each huge propeller disk was wider than *Sea Cliff*'s length. I'd never seen such a huge wreck, even during my dives on *Titanic*.

Yet this monster lay upside down, its bow and stern blasted off, smashed and impotent—a dinosaur overturned, never to right itself.

Vanity, I thought, echoing a passage of Christian scripture. All is vanity. Around me, I felt the chill presence of the dead young sailors who had fought and died in these bloody waters.

After the Solomon Island naval battles, the aircraft carrier became the dominant naval surface weapon during the bitter struggle between America and Japan. Aircraft carriers had evolved quickly between the two world wars. The short flightdecks improvised on hulls originally laid down as cruisers gave way to large fleet carriers specifically built to carry up to 50 torpedo planes or dive-bombers as well as several squadrons of fighters to defend against enemy planes. Indeed, the ordnance hurled into combat by a carrier's bombers superseded the battleship's longest-range cannon. A "battle wagon" such as the *Washington* could strike ships 15 or 20 miles way; the range of carrier-launched planes was several *hundred* miles.

The most decisive test of the aircraft carrier took place in the empty blue expanse of the Pacific northwest of tiny Midway Island in 1942.

In May of that year, a brilliant, eccentric U.S. Navy commander, Joseph Rochefort, shuffled among cluttered desks in a secret combat intelligence unit at the U.S. Navy's Pearl Harbor base. Rochefort had haunted Station Hypo for days on end since the Japanese air attack on December 7, 1941. He seldom left his basement and had taken to wearing bedroom slippers, with a musty flannel bathrobe over his rumpled khakis. His obsession was breaking the enemy JN 25 naval code, a tangle of over 45,000 five-digit units representing words or phrases. Some of Rochefort's code-breakers were former Navy bandsmen. Their musical talents, he reasoned, made them sensitive to the invisible logic woven into the enemy code.

On May 8, Rochefort's team deciphered messages indicating that the Japanese planned a large operation in the central Pacific. He became convinced their objective was Midway, a postage-stamp atoll whose airfield was America's westernmost outpost. But he had no proof. The decrypted radio traffic simply referred to "AF" as the target. To convince Washington that the island was threatened, Rochefort launched a cunning deception. The base on Midway transmitted an uncoded message, urgently requesting a water tanker because the distillation plant had "failed." Less than two days later, Station Hypo decoded a Japanese report that AF was short of water.

Not only had Rochefort learned the enemy's intentions, he was able to inform Admiral Chester Nimitz, Commander of the Pacific Fleet, that the massive enemy force was lead by aircraft carriers. Nimitz did not intend the U.S. Navy to be caught flat-footed as they had been at Pearl Harbor.

But Admiral Isoroku Yamamoto, the Japanese commander-in-chief, was equally determined. For months, his forces had swept across the Western Pacific and Southeast Asia. That May, the Imperial Navy launched an invasion force toward Port Moresby in New Guinea, intending to control the Coral Sea and cut the sea lanes between Australia and America. Nimitz stopped him during the crucial Battle of the Coral Sea, which was the first naval engagement fought exclusively between airplanes and ships. Japanese losses included scores of aircraft and a carrier. But they sank the American carrier *Lexington* and scored a direct hit on the *Yorktown*'s flightdeck with a 550-pound bomb.

Now *Lexington* was gone and *Yorktown* severely damaged just as the Japanese were preparing to attack Midway. The huge Japanese *kido butai* (strike force) would be spearheaded by four heavy and four light aircraft carriers capable of launching hundreds of fighters and bombers. As Nimitz planned his defense, he ordered the *Yorktown* to be

Following pages: Sea Cliff **approaches the giant Japanese battleship *Kirishima*, lying upside down on the floor of Iron Bottom Sound. The sinking ship turned turtle, its bow and stern breaking loose.**

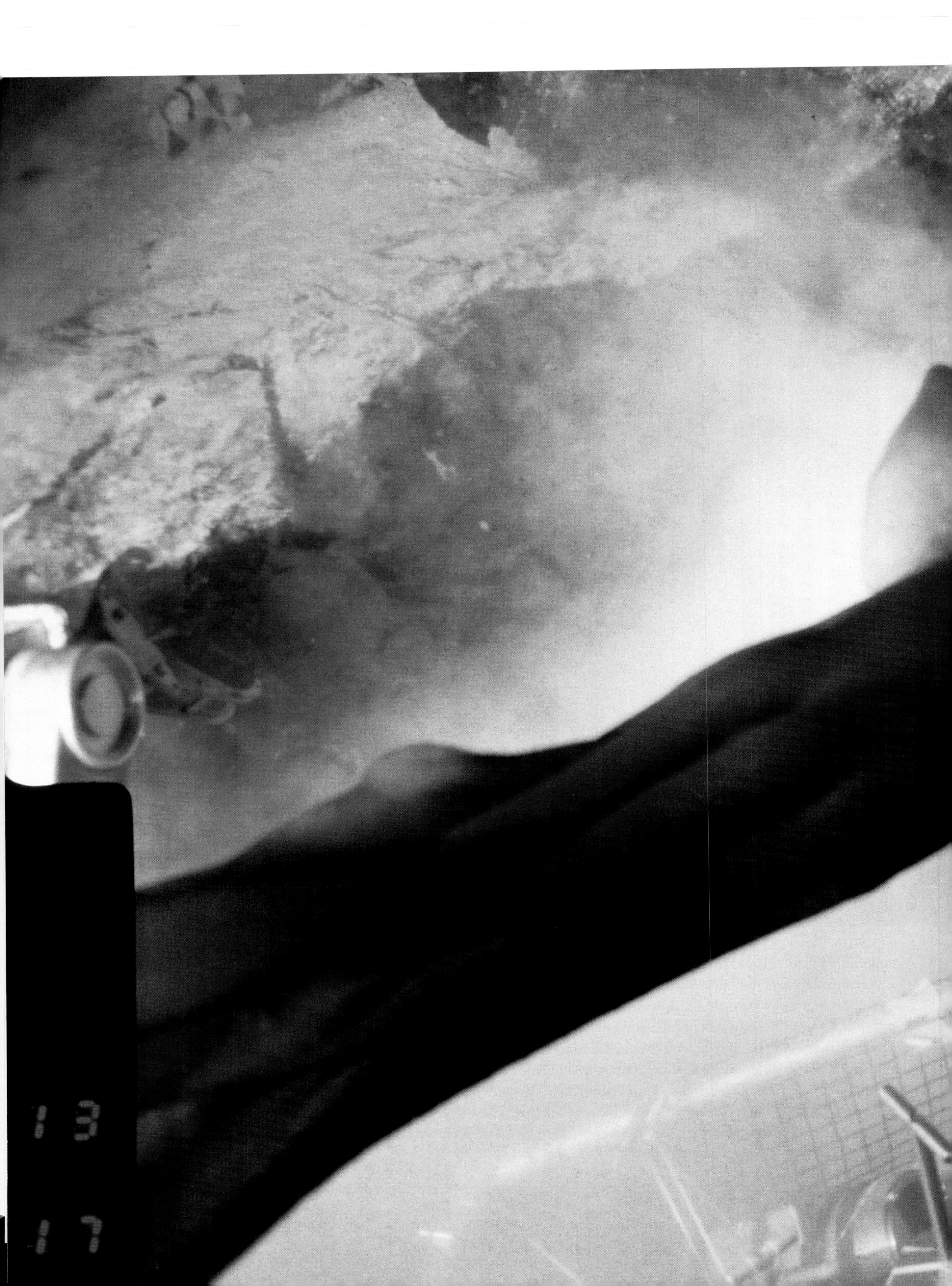

VII Science Beneath the Sea

Preceding pages: Like an eared apparition, a rare cirrate octopus passes *Alvin* 12,000 feet deep in the Caribbean's Cayman Trench. Such dives gave scientists their first look at rock layers deep within the Earth's crust.

On November 21, 1916, *Titanic*'s younger sister, *Britannic*, serving as a World War I hospital ship, struck a German mine and sank near the Greek island of Kea. The use of mines in shipping lanes symbolized the increasing cruelty of modern warfare.

My stomach growled with hunger as I lay with my face close to the *NR-1*'s starboard viewport. We moved silently, 450 feet below the sunny Aegean near the Greek island of Kea. The submarine's sonar confirmed that the wreck of H.M.H.S. *Britannic* lay just ahead in the slate blue twilight. It was August 1995, 10 years after the discovery of the vessel's older sister, R.M.S. *Titanic*.

The shadowgraph that the sub's skipper, Lieutenant Commander Dave Olivier, used to plan this dive had revealed *Britannic* lay on her starboard side, its 50,000-ton hull nearly intact, except for the bow forward of the bridge, which had been wrenched loose by the force of impact with the rocky bottom. Although a narrow gash separated the two sections, the sonar image displayed details of the ship's distinctive profile, including the multiple curled fingers of towering gantry davits that each had launched six lifeboats the day the ship sank, November 21, 1916.

In the control station above my observer's couch, Dave's crew edged us gingerly toward the shipwreck. They piloted the *NR-1* with a computerized "fly-by-wire" joystick to manipulate the pairs of tube-like forward and aft thrusters. The seafloor below glowed lime-white as Dave flipped on the intense greenish thallium iodide floodlights in the bow, which penetrated water better than conventional lighting.

We were at the hazardous stage of the first day's exploration. In shallow water, a long wreck like *Britannic* should have snagged miles of fishnet over the previous eight decades. None of us needed reminding that the submersible *Delta* had been trapped in a spider web during the exploration of *Lusitania* before jettisoning her propeller. But I had confidence in Olivier, his crew, and this remarkable vessel. Although Dave would conduct a thorough close-range sonar and visual reconnaissance before closing on *Britannic*'s hull, I knew *NR-1*'s powerful nuclear-electric drive and reserve buoyancy could best any fishnets that might ensnare us.

As we maneuvered closer, my stomach rumbled again. It was almost noon and I could smell the lasagna cooking in the sub's convection oven. But lunch would have to wait.

"Getting some good visual now," Dave said in my headset. The floodlights cut the darkness, revealing a long expanse of weedy riveted plate that moved slowly past my viewport as the crew expertly traversed the length of the hull. Perhaps because the ship lay on one side, and the four big funnels had fallen free on sinking, the wreck was free of fishnet. I was elated to see our shadowgraph confirmed; this was like having *Titanic* in one piece instead of the two widely separated sections we'd discovered in 1985.

Britannic had been launched in February 1914 from the same Belfast yard where her two older sister ships had been built. The White Star Line had originally intended to name her *Gigantic*, but the tragic night in 1912 among the North Atlantic ice fields had made that choice inappropriate. The name *Britannic*, however, had a patriotic ring that the line hoped would help restore public confidence after *Titanic*'s shocking loss. Indeed, the new ship was fitted with many innovations intended to make sinking extremely unlikely. Two-thirds of its hull, including boiler rooms and engines, was enclosed within an inner skin of steel plate. Its watertight bulkheads, which had proven too shallow on *Titanic*, extended well above sea level. The doors in these bulkheads could be remotely closed by electric motor and the compartments pumped dry. These were prudent precautions.

In 1995, we conducted a high-tech exploration of *Britannic* using two ROVs, *Voyager* and *Phantom*. We also inspected the wreck at close hand from the Navy's research submersible *NR-1*. A sonar graph produced amazing details of the wreck.

The painting shows the mine damage and the shattered bow, which collapsed at the high-speed impact with the stony bottom. Like the *Lusitania*, this wreck was shrouded with fishnets, making close-in inspection hazardous.

When World War I broke out, it entered service as His Majesty's Hospital Ship *Britannic*. With her hull painted bright white and emblazoned with big red crosses, the ship made regular runs between Great Britain and the eastern Mediterranean. The route took it to the Greek island of Lemnos in the northern Aegean, the staging area for sick and wounded Allied servicemen fighting in the Balkans, Palestine, and Mesopotamia. *Britannic* could carry over 3,300 patients for the fast voyage back to England.

In November 1916, it was on her sixth trip to Lemnos under the command of veteran White Star Captain Charles Bartlett, with 1,066 medical personnel and crew aboard. Unlike the *Titanic*'s Captain Smith, Bartlett had always gone to great lengths to avoid ice, and in wartime he made sure the ship's Red Cross hull markings were always freshly painted, although he didn't expect a German U-boat would torpedo a hospital ship.

But torpedoes were not the only hazard Bartlett faced. In October 1916, Kapitanleutnant Gustav Siess, commanding the *U-73*, had scouted the Kea channel near Athens, noting Allied ships passing close inshore. He laid 12 anchored mines, which floated near the surface across the shipping lane.

On the bright Tuesday morning of November 21, *Britannic* steamed through the channel, close enough to shore that the people on deck could gaze at the whitewashed villages among the island's hilly olive groves. There was a ripping crack from the starboard bow. The ship had struck one of Siess's mines, which ripped open the hull where a watertight bulkhead divided the second and third cargo holds, rapidly flooding both compartments and an adjacent boiler room.

Bartlett ordered all watertight doors shut electrically, but several failed, and water from the flooding forward holds surged aft. Within minutes, six compartments were flooded. Portholes that had been opened that morning to air hospital dormitories now slipped beneath the surface. Like the captain of the *Lusitania*, Bartlett instinctively headed at full speed toward the shallows of the nearby island, hoping to ground his vessel. But the bow-down list to starboard only worsened as the open portholes channeled water into the hull. Soon, it was time to abandon ship. Despite a tragic mishap in which two lifeboats were slashed apart by the still-turning propellers, all but 30 of those on board survived. Like *Titanic*, the ship rose to almost vertical before slipping beneath the calm water. It was 9:07 a.m., less than 30 minutes after *Britannic* had struck the mine.

Pioneering undersea explorer Jacques Cousteau used a two-man submersible called *Denise* to explore depths of up to 1,000 feet. *Denise*, here being lowered into the water from the deck of *Calypso,* had the ability to hover anywhere within its depth range.

The ship lay undisturbed two miles off the coast of Kea for the next 60 years. In 1976, French undersea explorer Jacques Cousteau surveyed the wreck, which his good friend Dr. Harold Carlson of MIT had located with side-scan sonar. Cousteau used his submersible *Soucoupe* to film the ship. The gallant Frenchman even took one of *Britannic*'s nurses on board the diving saucer to revisit the once proud vessel.

During our 1995 expedition, we acquired high-quality still pictures and videotape of *Britannic* using two ROVs, *Voyager* from Perry Tritech, Ltd. and *Phantom* from the National Undersea Research Program at the University of Connecticut. Our team also made several more dives in *NR-1* for close-in inspections of the wreck.

NR-1 was a true submarine—not simply a submersible vessel dependent on the atmosphere—that could operate completely independent of the surface for four weeks. I knew the sub well from a three-week scouting trip of the undersea Reykjanes Ridge near Iceland in 1984. Its amazingly compact nuclear reactor, the size of a lead-shielded 55-gallon oil drum, transformed the matter of fissionable uranium fuel into energy, providing an almost limitless power source. This was converted into electricity to drive the boat's main propellers, thrusters, and auxiliary systems, including a seawater desalinization plant. Our oxygen came from canisters that were "cooked" in a small furnace whenever automated monitoring equipment detected our breathing air needed more of that vital gas. The life support system also monitored carbon dioxide levels and automatically scrubbed the air before a dangerous buildup occurred.

We worked around the clock in rotating six-hour watches, which allowed the crew the dubious comfort of sharing a "hot bunk" with the man who had just left his berth to go back on duty. The crew jury-rigged me a pipe bunk wedged like a cupboard shelf between racks of instruments in the shoulder-wide main passageway. With plenty of fresh water, each man did have a daily "shower." This involved dipping from a five-gallon bucket of warm water on a grating. After a good sponge bath, the rinse-off was pumped into a holding tank.

Meals were important, not just for nutrition, but also to punctuate the otherwise unbounded flow of time, which could be disorienting at times. We were operating down at 3,000 feet, silently gliding across giant terraces separated by steep volcanic scarps, far below the world where clocks coincided with daylight and night. So a platter of fried chicken from the sub's bountiful freezer reinforced the concept that you were sitting down to supper, even though it might be 3:00 a.m. in the world above the sub.

In those three weeks, *NR-1*'s skipper, Commander Ed Giambastiani, took us on a twisting 120-mile route across the complex seafloor terrain. Using the sub's unique maneuvering ability, we crept up the sides of active volcanic cones like a stealthy tank masking its advance beneath the crests of hills. Ed and his executive officer, Lieutenant Commander Charles Anderson, cautiously poked the sub's bulbous snout deep inside the glistening black walls of a wide lava tube. We slowly wove our way among jagged basalt pinnacles and through deep canyons. Above us, much closer to the surface, U.S. Navy attack submarines quietly shadowed Soviet missile boats entering the open Atlantic through the narrow Iceland-Faeroe "chokepoint." Those rival submariners no doubt saw their powerful nuclear vessels as swift, maneuverable undersea fighter planes. But their down-looking sonars would not have revealed our presence, showing only a complex, jumbled bottom in which we were invisible. From our hidden lair, the submarines passing above were more like lumbering blimps in the watery sky.

During that cruise I again remembered the impression that Jules Verne's visionary novel *Twenty Thousand Leagues Under the Sea* had made on me as a boy growing up near San Diego. When I launched my mason jar models of Captain Nemo's *Nautilus* in the tidal pools of Mission bay, I never dreamed that one day I would actually live and work aboard a vessel every bit as engrossing as that fictional submarine.

This interior view of *Soucoupe*, Saucer, shows the austere two-person crew compartment. Later, deep-diving submersibles were much more complex and normally carried a crew of three.

***Following pages:* Humans have been fascinated for centuries by what lies beneath the surface of the sea. Jacques Cousteau invented the Aqua-Lung in the 1940s, allowing people to explore shallow water as freely as fish.**

Humans have been fascinated by what lies beneath the glittering surface of the sea for thousands of years. We know that naked sponge divers pursued their hazardous trade in the ancient Mediterranean. In Japan and Polynesia, pearl divers holding their breath could reach depths down to 160 feet. But it wasn't until the invention of leather goggles with glass lenses that Mediterranean coral divers or pearl divers of the Pacific obtained a clear view of the underwater world. The innovation of watertight goggles was periodically forgotten and reinvented in the West, most recently by my fellow Californian Guy Gilpatric, who combined rubber goggles with shatterproof lenses and swim fins in the 1930s.

positive buoyancy. And finally, if all these steps didn't work, the forward section, which contained the pressure sphere, could be mechanically detached, leaving behind more than half the crippled sub. I was convinced that *Alvin* was the perfect tool for a hands-on geologist like me to take to the seafloor.

I got my chance to test the submersible's potential during the period of intellectual fervor over the revolutionary theory of plate tectonics—Tekton was the name of the carpenter in the *Iliad*—that gripped the Earth sciences in my final years as a graduate student. One winter night in 1968, I attended a lecture at the Massachusetts Institute of Technology by geology professor Patrick M. Hurley on the comparison among sedimentary rocks in the Americas, Western Europe, and West Africa. His conclusion was stunning: The strata of these formations were not merely similar, they were identical. The rock of the Appalachian Mountains matched that of the Channel Islands near the French coast, the northern British Isles, and Scandinavia. Almost certainly, the continents had once been joined, but were now separated by the 3,000-mile Atlantic. By comparing the rock of Brazil with West Africa's bulge, Hurley demonstrated that Africa and South America had also once been attached, the bulge of Brazil nestling into the West African Bight of Benin.

During the 1974 diving season on the Mid-Atlantic Ridge, Project FAMOUS made use of several submersibles, including *Alvin*. I have repeatedly descended in *Alvin*'s cramped sphere to the Mid-Ocean Ridge, including this descent to the Galápagos Rift three years later.

Hurley presented persuasive scientific evidence to prove the once discredited theory of continental drift, soon to be known as plate tectonics. This was literally Earth-shaking stuff, and the small world of professional geology exploded in an uproar. Classical geology did not endorse movable continents: They rise and fall vertically as their rock is eroded away to form thick sedimentary deposits whose weight then forces up immense adjoining blocks.

But the advocates of plate tectonics posited that continents rest on vast, relatively light rafts of rocky crustal plate, which drift on a spherical pond of denser superhot mantle. Crustal plates are always separating, colliding, or grinding alongside each other at fault lines. When two plates such as the North American and European pull apart, molten magma rises along an expanding raised seam like the Mid-Atlantic Ridge, part of a longer, planet-spanning undersea mountain range. At collision points, the titanic warping of the crustal plates can create deep ocean trenches or thrust up mountains such as the Andes and Himalaya.

By the mid-1970s, the earth sciences had generally come to accept the mounting evidence of plate tectonics, as startling as it was. But there the agreement within the science establishment ended. Geophysicists, who studied such fundamental phenomena as gravity, magnetism, and naturally occurring radioactivity, felt they should be the pioneers on the frontier of plate tectonics. Geologists like me, whom many geophysicists dismissed as technicians, believed we should receive most of the increasingly scarce research funding to prove the new theory. We studied rock, the stuff the crustal plates were made of.

And the seafloor, specifically the Mid-Atlantic Ridge, seemed the ideal place for geologists to conduct their field trips because the evidence of plate tectonics would be exposed to clear view. Researchers towing instruments in the 1950s had found intriguing bands of zebra-stripe magnetic patterns on either side of the Ridge. Like the rock Hurley had studied, these patterns had the same width and sequence both east and west of the mid-sea mountains. Plate tectonics advocates believed iron-rich basalt had extruded as one molten mass, then spread in parallel bands. The zebra pattern was due to the periodic reversal of the Earth's magnetic poles over millions of years, a record that was "frozen" into this basalt as it cooled.

Geologists were eager to dive on the Ridge, 9,000 feet below the mid-Atlantic, conduct field observations, and collect samples. Geophysicists wanted to continue studying the Ridge from surface vessels with magnetometers, gravimeters, and seismic sounding equipment.

We reached a compromise. The French-American Mid-Ocean Undersea Study (FAMOUS) of 1973-74 combined traditional geophysical surface probes, photo surveys with deep-towed camera sleds, and bottom investigation from manned submersibles. The French Navy's bathyscaphe *Archimede* made the initial dives in 1973. But the bulk of the manned diving came during the 1974 season, using *Archimede* and the smaller French deep submersible *Cyana*, as well as *Alvin*, which had received a new titanium pressure sphere allowing it to descend more than 10,000 feet. Much of the diving took place on the central rift valley of the Mid-Atlantic Ridge, about 400 miles southwest of the Azores Islands, where it was theorized the North American and the African crustal plates were slowly spreading apart.

On July 1, 1974, the morning after my 30th birthday, I slipped across the gangplank connecting *Alvin* to its support twin-pontoon ship *Lulu* and mounted the submersible's sail, volcanologist Jim Moore right behind me. We dropped into the narrow, instrument-crammed pressure sphere below, where our pilot, Jack Donnelly, joined us.

Our principal investigation site, about 9,000 feet below, was a twisting twin ridge divided by a wide central rift valley of wild volcanic wasteland. A sheer, 1,000-foot west wall formed the rampart of the North American crustal plate. West of this serpentine cliff, the expanding seafloor was a chaos of volcanic domes, fissures, and lower cliffs. The central rift valley was broken by linear gullies. Two conical undersea volcanoes, Mount Venus to the north and nearby Mount Pluto, dominated this section of the Ridge.

Camera-sled reconnaissance including photos from WHOI's new Acoustically Navigated Geological Undersea Surveyor (*ANGUS*) revealed fresh-looking lava extrusions, strong evidence that the crustal plates were separating and new magma from the mantle below was surging up to fill the widening gap. But the rift was still narrow enough for *Alvin* to maneuver from one plate to the other on a single dive.

"Oxygen on, blower running," Jack Donnelly read from his checklist. He carefully examined the vital hatch seal and locking mechanism.

"*Lulu*," Jack radioed, "my hatch is closed. My tracking pinger and underwater phone are on."

I quashed a momentary flutter of nerves, remembering the previous summer when I'd almost suffocated aboard the French bathyscaphe *Archimede*. But this little submersible was practically brand new after its recent overhaul. I was ready.

"Request permission to dive," Jack called.

"Roger, *Alvin*," fellow pilot Dudley Foster radioed back from the support ship. "You are clear to dive. Present water depth is 8,700 feet. Good luck."

We fell at one foot a second through the warm blue surface water. Quickly the color outside the viewport faded into indigo. The last chill afterglow of sunlight vanished.

"*Alvin*," Dudley's echoing voice sounded in the acoustic phone speaker. "Your present position is X 55.6, Y 100.4. We suggest you drive a course of 180 degrees."

The surface navigator followed our position by triangulating *Alvin*'s sonar pinger within the anchored transponder network floating 900 feet above the bottom. The system normally worked perfectly, but the rough terrain of the Ridge could produce false sonar echoes. Then there was the "flipper factor": Our little pals, the dolphins, were intrigued by the transponders and quickly learned to mimic their frequencies.

But today, our navigation was right on the money. At a depth of 8,500 feet, the short-range sonar pinged with a solid contact. Jack dropped descent weights to neutralize our buoyancy. Soon our floodlights revealed a dull bottom glow.

"I have visual," I sang out from my viewport, sighting a black lava cliff face. After Jack rested *Alvin*'s bottom skid neatly against the steep formation, Jim and I used the manipulator arm to take samples and place them in the external collection tray. The lava was so fresh here that it was not dusted with the gray manganese deposits that coated rocks beyond the central rift after years of exposure to seawater, persuasive evidence that tectonic seafloor spreading was still in progress. We investigated all the features shown in the *ANGUS* reconnaissance photos: twisting, foot-thick toothpaste cylinders; pillow lava, looking like dirty foot stools; and domed haystack formations larger than *Alvin*. Jim and I dictated field notes nonstop into our recorders and obtained a visual record of the survey with the exterior video camera and the twin-camera stereoscopic system. Almost two miles below the surface, we were practicing some pretty sophisticated geology.

Near the end of the dive, we encountered an unmistakable tectonic feature: a knife-edge fracture running toward the south.

"Look at that fault line," Jim said. "It's clearly linear."

A near-fatal catastrophe occurred on the Mid-Atlantic Ridge in 1974 when *Alvin* pilot Jack Donnelly, with geologists Bill Brian and Jim Moore on board, accidentally wedged the submersible (far left) in a narrow fissure. Only after great effort did Donnelly free *Alvin*. He signals his safe return as he debarks to the mother ship *Lulu*.

The fissure cut the glassy lava like a sword stroke. I'd seen such linear fractures on California's San Andreas Fault, where the North American and Pacific crustal plates abutted. Here we had dramatic evidence of the tectonic process taking place outside *Alvin*'s viewports. As Jack lifted us back to the world of sunlight, I knew I'd earned my pay that day as both an explorer and a scientist.

But we learned that the manned submersible was not glitch-free when Jack Donnelly piloted geologists Bill Bryan and Jim Moore into the rift valley to investigate deeper fissures. They were probing for seawater that might have percolated down to magma chambers, then surged back up through rifts to form hot springs.

As their navigator in the cramped control van aboard *Lulu*, I was busiest when *Alvin* drove along the bottom between survey stations. Two hours into the dive, I

A skin diver gives *Alvin*'s exterior a safety inspection prior to a dive on the Mid-Atlantic Ridge. The mesh basket at the prow was vital to our research and often returned to the surface carrying unique geological samples.

saw that the plotting pen marking their position was bouncing continuously on the same spot of the plotting sheet. Why hadn't Jack moved on schedule?

"*Alvin*," I called, on the sound phone, "are you still at station four?" I was trying to nudge them along before mission time elapsed.

My weird Donald Duck echo bounced from the bottom 9,000 feet below, followed by Donnelly's strained voice. "We're trying. But we don't seem to be able to rise."

In edgy tones, Jack explained that they'd entered a deep fissure from which warmer water was rising. *Alvin* had ventured inside, nose angled slightly down, its temperature sensor sniffing like a retriever's snout. The two geologists were captivated by the size of the fissure. Jack concentrated on his controls. None of them saw the upper walls closing above.

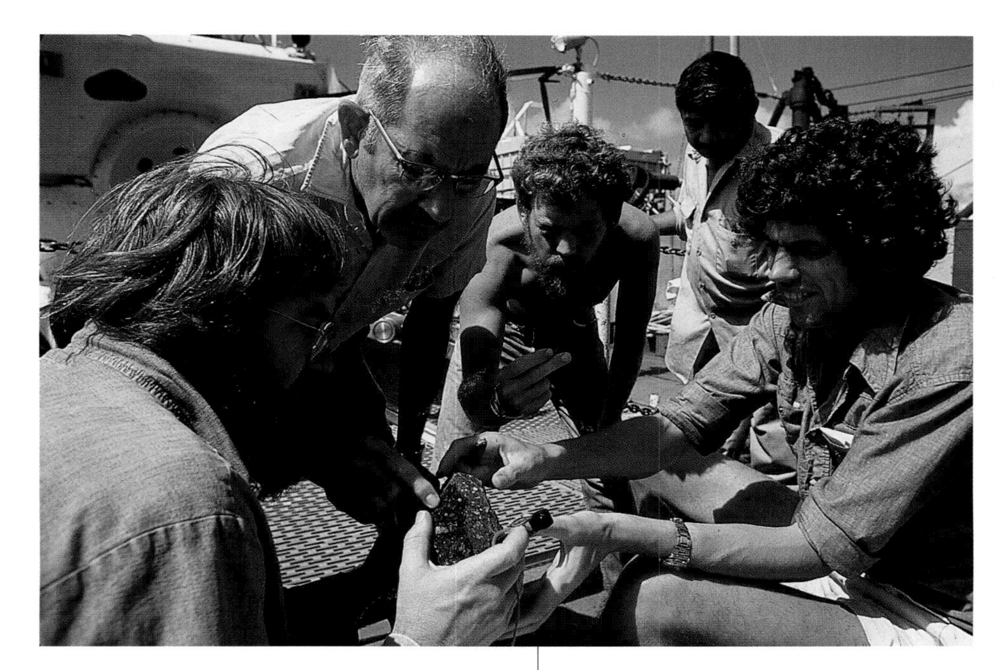

During our 1975 expedition to the Cayman Trench, we recovered intriguing rocks that helped us better understand the internal composition of the Earth's crust. Here my mentor, Professor Kenneth O. Emery (center), and other scientists inspect a sample that *Alvin* has returned from the deep.

"Can you get Val?" Jack called, asking for Val Wilson, *Alvin*'s senior pilot. "I need to talk to him."

The normal banter in the van fell silent. Val sat beside me at the navigation plot speaking into the mike in a clipped monotone as he methodically asked Jack to turn the lift props first in one direction, then in another, while simultaneously applying spurts of power ahead and astern.

For 90 long minutes, Jack Donnelly's voice warbled up from the depths. No combination of maneuvers would free *Alvin* from the lava mousetrap. The submersible had life support reserves for three days. But if it were hopelessly stuck in that narrow crack, even the other expedition submersibles, the French Navy's nimble *Cyana* and their lumbering bathyscaphe *Archimede,* could not help it in that time. And the ultimate emergency procedure, detaching the sphere, would simply doom the crew. Without propellers, the sphere with its attached buoyant hull section would pop to the top of the overhanging fissure and remain trapped forever.

Two hours and ten minutes after the emergency began, Jack announced, "We're clear and under way again to our next station." Deft coordination of propeller thrusts and ballast shifts had finally freed them. At sunset, Jack slid *Alvin* into its slipway between *Lulu*'s pontoons. Inspecting the submersible, we found crumbs of black lava in cracks along the hull.

"That was like backing a Cadillac out of a VW parking space," was all Jack said.

In June 1977, I dove in the Cayman Trough aboard the Navy's bathyscaphe *Trieste II*. This undersea chasm marked a tectonic rift where plates were grinding past each other as in the San Andreas Fault, but a right turn in this fault led to a divergence, forming a gargantuan hole or trough five times deeper than the Grand Canyon. Lieutenant Commander. Kurt Newell and his co-pilot Chief Petty Officer George Ellis took me on our deepest dive, more than 20,000 feet down the side of an endlessly plunging lava canyon.

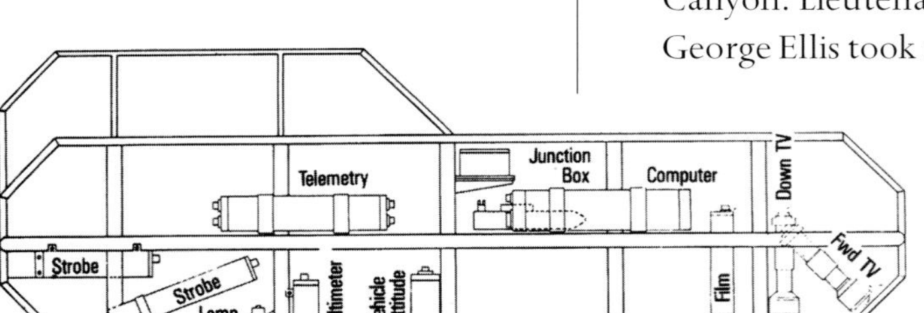

The vehicle ***ANGUS***, which was first used in Project FAMOUS, was later replaced by the more sophisticated ***Argo***, illustrated above.

After dozing during the five-hour descent, I noticed the first shadowy trace of the bottom on the echo sounder. Peering through the thick oval viewport, all I saw was the floodlit glare of slowly drifting marine snow that might have been falling from the distant surface for weeks on end. Then George began to chant the diminishing altitude to the bottom: "Five hundred feet...four hundred feet...three hundred...coming down, plenty of room...."

Plunging at terminal velocity, *Trieste II* was vulnerable to violent bottom impact, which might damage our thin-sided gasoline buoyancy tanks, so we had to be ready to release steel shot ballast. I strained to see any hint of solid structure.

"Two hundred feet," George called.

Then I glimpsed a dark rocky slope.

"Bottom!" I yelled. "It's coming up fast."

"Impossible," Kurt retorted. "The sounder reads one hundred and fifty feet."

"I don't care what it says. I can see the bottom."

Kurt Newell believed me. He began dumping steel shot. But our tons of mass carried great momentum. The front of *Trieste II*'s flotation tank slammed into the cliff I had seen moments before. We bounced down the slope of a steep volcanic cone. Holding my breath, I watched the gray steel hull snout plow through the lava flow, churning up clouds of pulverized rock. Then I saw a ripply wave of polychrome bubbles. "Gasoline in the water," I croaked as calmly as I could.

The rainbow smear of gasoline meant the flotation tank had ruptured. As we watched in stricken horror, our buoyancy was hemorrhaging. We might never reach the surface again.

Kurt and George frantically performed the emergency jettison maneuver, dumping all the remaining ballast. None of us spoke. We only gazed at the blinking digits of the metric ascent-rate display. If we had enough residual positive buoyancy in our ruptured flotation tanks, we would rise at about 30 meters per minute. If not, we were dead.

The display fluttered: 31, 29, 30, 28, 27, 31...30.

I closed my eyes and breathed slowly to dampen the useless adrenaline. Finally, the winking red pulse steadied: 29, 29, 28...31, 30.... *Trieste II* had positive buoyancy.

We broke the surface in the last peach glow of the Caribbean sunset. Gripping the wet edge of the tower as I climbed from the float tunnel, I made a vow: Never again will I dive in a bathyscaphe. The electrical fire in *Archimede* had nearly choked me that first season on the Mid-Atlantic Ridge, and today *Trieste II* had almost finished the job. There simply had to be a better way to explore the deep ocean than in these dinosaurs.

Although best known for its discovery of the *Titanic*, *Argo* continues in *ANGUS*'s footsteps in mapping the Mid-Ocean Ridge.

I was on sabbatical from WHOI, working on plate tectonic research at Stanford, when I conceived of the means to carry explorers' questing senses and gripping hands to the seafloor while protecting their fragile bodies on the surface.

My Palo Alto study was stacked with trays of color slides from deep-sea geological expeditions, yet I had actually covered less than 50 miles along the 40,000-mile mountain range that crossed the planet beneath the oceans. A rugged little submersible like *Alvin* was just too slow and too dependent on the surface to serve as an exploration tool.

As I wrote my academic papers that year, I became convinced that the day of manned submersibles in ocean exploration had already peaked. Even our old towed camera sled, *ANGUS*—basically a welded pipe frame holding two 35-mm cameras with extended film magazines and strobe lights—could cover more bottom terrain in 24 hours than *Alvin* could in a month. But *ANGUS* was a "dope on a rope," as we'd taken to calling it. The still pictures it produced were often useful, but had to be processed in a lab aboard the surface ship, a process that took hours. If *ANGUS* found interesting bottom features, the ship had to double back and try, sometimes in vain, to find the intriguing site again. I knew from personal experience just how frustrating this could be.

I wanted a system that would combine the tireless search endurance of a towed camera sled with the flexibility of a manned submersible, in which explorer-scientists directly observed in "real time" through viewports and collected samples with the manipulator arm. I was still mulling over this challenging concept when I read about the work of researchers in nearby Silicon Valley, the heart of America's high technology frontier. Their amazing innovations in computer chips and low-light

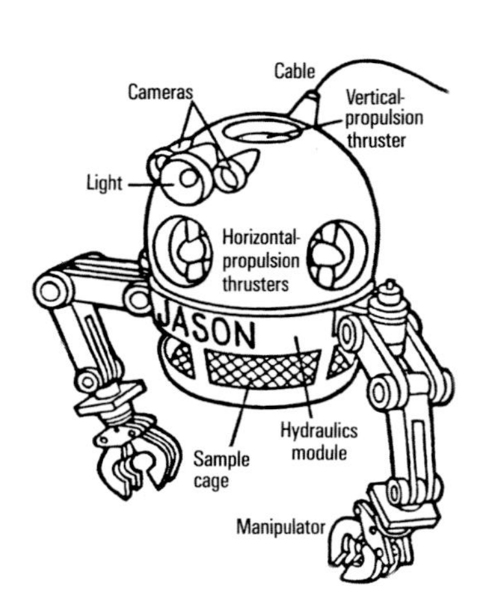

This early concept of the ROV *Jason* (above) incorporates the actual vehicle's essential elements. We originally assumed the ROV would need a streamlined, spherical structure, but the actual *Jason* (right) has a far different form, with more lights and imaging systems, and only one manipulator arm. With its multiple thrusters, *Jason* is a nimble acrobat.

television might provide the robotic senses I was seeking. And fiber-optic cables they were developing would be a quantum leap beyond *ANGUS*'s commercial trawl cable "rope" or even the more sophisticated coaxial cable the Scripps Oceanographic Institute's *Deep-Tow* used.

Sketching on notebook paper, I envisioned a towed vehicle connected by fiber-optic cable to a control van on the surface ship. This multichannel umbilical would carry crisp color television and sonar images and also transmit remote operational commands to the thrusters, sensors, cameras, and robotic grasping arms. In effect, we would have a miniature *Alvin* at the end of a tether, capable of extended deep exploration without concerns about crew life support or having to return each day after only three or four hours on the bottom.

The "crew" of scientists would work in shirtsleeve comfort, but they would be virtually present on the bottom far below, instantaneously observing phenomena on video monitors and sonar shadowgraphs. I named this ability to be simultaneously in the surface control van and on the seafloor "telepresence."

One morning I awoke at dawn and went to my study. Before falling to sleep, I'd been reading Greek mythology. Now I knew what I'd call my new system: *Argo-Jason*, in honor of the first mythical explorer of Western civilization.

Argo would be the exploration trailblazer, a sophisticated towed video- and still-camera sled augmented by ultra-sensitive side-scan sonar. *Jason* would be a remotely operated vehicle that would closely investigate and robotically sample interesting geological, biological, and archaeological sites that *Argo* had discovered. Propelled by its own thrusters, *Jason* would not need the surface ship to tow it across a site. The two systems would operate from research vessels equipped with dynamic positioning systems that allowed them to remain stationary above the ocean bottom or to slowly move along precise search tracks.

I managed to convince my friend Sam Matthews at NATIONAL GEOGRAPHIC to carry an illustration of this pipe-dream vehicle in an article he wrote on the future of oceanography. Dr. Ballard's *Argo* and *Jason*, he said, were in the "design and prototype stage." This was academic-explorer gamesmanship, bread on the water that might attract deep-pocket support. Luckily for me, my friends at ONR dug into one of their pockets and anteed up some of the preliminary funding. While my group at Woods Hole was testing the initial technology, I pitched the concept to Navy Secretary John Lehman and Admiral Ron Thunman. The Navy came through with multi-year research and development support. I formed the Deep Submergence Laboratory at Woods Hole to build my new system. *Argo-Jason* began its voyage from the realm of myth to scientific reality. Years later that fully mature system would show its true value on important underwater archaeological sites, most notably during the recovery of delicate Iron Age terra-cotta artifacts from *Tanit* and *Elissa* off Ashkelon.

On the deck of a ship, *Jason* was an ugly little fellow, squat and bug-eyed, a ton-and-a-half block of plumbing and jutting floodlights. Submerged in its element, however, *Jason* was a graceful, unflagging athlete. It could pirouette precisely, or hold station for hours, automatically controlled with its seven thruster propellers. If need be, it could enter a dangerous cul-de-sac such as the Mid-Atlantic Ridge fissure that had almost trapped *Alvin*.

And today, more than 20 years after I made those tentative first sketches in my Palo Alto study, I have the satisfaction of knowing underwater exploration has come a long way from Alexander's legendary diving bell.

VIII The Living Ocean

Preceding pages: **An albino fish swims among large red tube worms around a hydrothermal vent, which we discovered on the Galápagos Rift in 1977. Such bizarre seafloor creatures exist on a food chain based on the chemosynthesis created in the vents, rather than on the photosynthesis previously thought to be the basis of all life on Earth.**

On February 15, 1977, the *Knorr* cruised at two knots through the torrid equatorial night, 9,000 feet above the Galápagos Rift in the eastern Pacific. Our expedition's research ship was 200 miles north of the Galápagos Islands, where the persistently seasick Charles Darwin had gratefully stumbled ashore in September 1835 and forever changed human understanding of life on Earth. The seafloor below was a much faster tectonic spreading zone than the Mid-Atlantic Ridge that we had studied in 1974. Here, the Cocos and Nazca Plates were separating at a speed of six centimeters a year. The resulting upsurge of molten magma from below the crust had created a sea-bottom highland of pillow and mound lava slashed by angular stress cracks.

As our rugged old *ANGUS* glided through the chill darkness of the abyss just 12 feet above the lava, strobes flashed every 10 seconds to capture another photo in an overlapping 3000-image mosaic. A thermistor on the camera sled sniffed for changes in water temperature. Bach had replaced Merle Haggard on the control van stereo and Earl Young had just relieved Al Driscoll at the flyer's station. My strong right hand, Cathy Offinger, was at the navigator's position, where she plotted both *Knorr* and *ANGUS*'s coordinates from the neon green pulses on the computer screen. I was going to stretch my legs on deck when Scripps Oceanographic Institute graduate student Kathy Crane lifted *ANGUS*'s telemetry scroll.

"Looks like a temperature anomaly," she said, pointing at the thermistor data, which had registered a sudden jump from the near freezing 2.5°C common to seafloors worldwide. I examined the unrolling scroll. Probably just a telemetry glitch, I thought. Our crude data transmittal system relied on acoustic signals rather than a more dependable hard-wire circuit. But this heat spike seemed too prolonged for a data garble.

"Maybe we *do* have an anomaly," I mused.

The crew in the ship's laboratory was suddenly all business, aware that we might have stumbled onto something interesting during our first run.

"Let's make sure to log all this, gang," Cathy Offinger reminded us.

Kathy Crane unfolded the scroll. "Anomaly begins at exactly nineteen-oh-nine," she said.

As if on cue, WHOI's Dick von Herzen, a geophysicist, and geochemist Jack Corliss of Oregon State entered the lab. Jack, Dick, and Stanford geologist Jerry van Andel were the people who had gotten the Galápagos Hydrothermal Expedition moving. After Project FAMOUS, Dick had zeroed in on the pictures we'd taken of the Mid-Atlantic Ridge fissures. He was a heat expert who'd cruised the world, dropping probes to the seafloor to measure thermal energy. Earlier, he had developed an elaborate model of heat flow from the Earth's super-hot core, up through the mantle, and into the crustal plates. Mid-ocean ridges, which were sculpted by heat, intrigued him. The structure of the Mid-Atlantic Ridge had initially seemed to fit Dick's theoretical cooling curve of tectonic spreading in the ocean crust perfectly.

But Dick admitted that the data we'd brought back from the Atlantic had ruined his theory. Although the Ridge's topography seemed ideal, the heat distribution did not. The temperature of the seafloor in the central rift valley was much lower than Dick's carefully constructed model predicted. We verified our data with him; they were not the problem.

"Something's got to be stealing heat from the ocean bottom, Bob," he'd told me.

The favored hypothesis was that hydrothermal circulation occurred in tectonic rift valleys: Frigid seawater seeped inside the fissured lava crust, was superheated down at the magma chambers, then gushed up as hot springs. In the process, thermal energy was "stolen."

Although *Alvin* had gotten stuck in that fissure looking for them, we'd found no undersea geysers on the floor of the Atlantic. Since the tectonic spreading in the Pacific was relatively so much faster, however, the magma chambers were constantly replenished with molten rock. We stood a good chance of finding hydrothermal circulation on a site like the Galápagos Rift. Unfortunately, sponsors weren't lined up to fund an expensive expedition to investigate this obscure phenomenon. The WHOI-Oregon State-Scripps-MIT alliance we had cobbled together finally received support from the National Science Foundation under the International Decade of Ocean Exploration program. When the cruise planned to satisfy the curiosity of a few earth scientists departed, we could never have guessed we'd make monumental biological discoveries.

The clams' flesh is deep red with hemoglobin-laden blood, an adaptation to the harsh conditions of the icy seafloor. This evolutionary tactic prevailed among the vent creatures we discovered.

In 1976, reconnaissance with the Scripps team's *Deep Tow* had produced a photograph of white clamshells—and a single brown beer bottle—strewn across the central axis of the Galápagos Rift. Since no one had ever seen clam colonies at that depth, the Scripps people had naturally surmised that this was garbage dumped from a passing ship and had called the site "Clambake."

I certainly wasn't thinking about clams a year later aboard *Knorr* as I studied the thermistor telemetry scroll. Three minutes after the anomaly began, the parallel lines rejoined. *ANGUS* had passed beyond the area of higher temperature. But with our crude instruments, we could not connect the temperature spike with a bottom feature until we finished the 12-hour run, winched *ANGUS* up on board, and developed the long roll of film.

I woke at dawn and went to the photo lab to join Jack Corliss, Dick von Herzen, and the *ANGUS* team. We leaned over the lab table as Pete Petrone, our NATIONAL GEOGRAPHIC photo technician, threaded the roll of film on the scanning projector. The screen blurred until he zeroed in on the color images spanning the three-minute temperature anomaly beginning at 19:00 hours.

"Clams!" Jack shouted.

We stared silently at big white clamshells jumbled so thickly together they almost hid the pillow lava to which they were attached.

"*Hundreds* of clams," I added, counting rapidly. "And look at the color of the water."

The image was turbid, almost milky, unlike the limpid water we'd always seen on fresh, sediment-free lava flows.

"This sure isn't some fisherman's garbage dump," I said.

The creatures were alive, robust, enjoying a happy clam's existence in the icy black abyssal desert. But their unlikely habitat was the supposedly near sterile void of the Galápagos Rift, not a sunlit bed of eelgrass in Chesapeake Bay. I had been on the sea bottom scores of times, but had never encountered such an abundance of life. On the smooth gray sediment of the Atlantic floor, I'd seen the occasional weed-like sea lily, a small primitive animal related to the brittle stars I had found once or twice in the Gulf of Maine. Often I'd see holothurians (sea cucumbers), or at least their tracks moving across the mud, and an occasional deep-dwelling shrimp, but as a whole, the abyssal bottom was a poor environment. Those few creatures that survived there eked out a precarious existence from a food chain anchored in the sediment, which had accumulated over millions of years from the marine snow drifting down from the surface.

Bottom-dwelling creatures were the isolated end-users of the photosynthesis that powered the trillions of tons of marine plankton in the upper ocean layers on which more complex life-forms depended. But because the seafloor was so poor in nutrients, the animals living there were generally small and widely separated. And I had certainly never seen any indigenous animal living on freshly extruded lava, which was generally free of sediment. Yet here we were, studying pictures of a healthy colony of large clams that had staked their claim on just such lava.

"What do you think's going on down there?" Jack Corliss asked. We were all wondering the same thing.

We didn't have long to wait to find some answers. The tender *Lulu* arrived on the site, carrying the submersible *Alvin* and Jerry van Andel. Now that *ANGUS* had located a colony of clams living around an area of warmer water, our expedition had suddenly expanded into the realm of biology.

On the morning of February 17, 1977, *Alvin* pilot Jack Donnelly took Corliss and van Andel down to investigate the mysterious clam colony. The bottom was an endless field of ripply black "pahoe-hoe" flow lava—glassy, soot-covered snowdrifts. Corliss was carefully checking the readings of the temperature sensor resting in *Alvin*'s sample basket because the two scientists wanted to assemble a careful heat curve as they approached the clam colony.

Donnelly expertly inched *Alvin* forward as Corliss listened for beeps on his instrument that would register each thousandth of a degree Celsius in temperature increase. The instrument remained silent as they passed from the rippled lava into a circle of shiny new domed pillows. Then they glided over some small fissures and the pressure sphere was filled with the incessant beeping of the temperature probe.

"We can see the clams," Jerry van Andel called up on the acoustic phone.

Alvin's powerful floodlights illuminated a bizarre oasis of life, a 60-foot (20- meter) circle completely separate from the sterile nearby lava. Clams jutted in dense, overlapping ranks, their shells agape as they passed seawater through their bivalve digestive system. Donnelly used *Alvin*'s mechanical arm to insert the heat sensor into the larger fissures. The temperature monitor trilled until the blinking digits steadied at 16°C (61°F), warmer than the rich clam beds off New England's rocky shore.

"We're sampling a hydrothermal vent," Corliss announced.

Up on *Lulu*, Jack's graduate student Debra Stakes operated the acoustic phone link. She listened as he described the milky blue billows shimmering in curtains

Like oases, some vents play host to a dominant species such as tube worms (left) or clams, as well as dependent creatures such as crabs and lower invertebrates.

from the fissures. The color suggested dissolved manganese particles and other unknown minerals.

Then Jack asked the question that concerned us all, "Debra, isn't the deep ocean supposed to be like a desert?"

We were earth science people, not marine biologists. "Yes..." she answered tentatively. "That's what I've always been taught."

"Well," Jack Corliss continued, "there's all these *animals* down here."

He carefully described the totally unexpected scene outside the viewports. Oval and about as large as the chalk-white china plates in highway diners, the clams

Alvin photographs a cleaning crew of albino crabs living on the perimeter of an active colony of large mussels (center) at a hydrothermal vent on the Galápagos Rift.

actually had the appearance of dishes stacked densely in a drying rack. Spindly white crabs picked their way daintily among the clams. There were also albino lobsters as "chunky" as a rolled newspaper. Then there were "really weird" creatures that looked like "swollen orange dandelions."

"Ask him to take a lot of pictures," I radioed Debra aboard *Lulu*.

In the main expedition lab on *Knorr*, we followed this dive with increasing interest and incredulity. Before *Alvin* reached the surface at sunset, they had visited two more hydrothermal vents, each hosting an oasis teeming with life, along a linear mound in the Rift's central valley. In the submersible's specimen tray were several cocoa-brown mussels, a big clam with beef-red flesh, and some small blocks of lava coated with a beige scum that might be a mat of colonial bacteria.

Over the coming days, *Alvin* retrieved more specimens and several liters of sea water from the hydrothermal vents. MIT chemist John Edmond analyzed the water samples while we were at dinner in the mess hall, but he didn't need very sophisticated techniques to draw us to the lab. The unmistakable rotten-eggs stench of hydrogen sulfide quickly spread through the ship. The water swirling around the clam colonies contained enough of this dissolved gas to form a lethal environment on land; yet, life was diverse and abundant down at the site we now called "Clambake I."

That night we held an old-fashioned grad school bull session trying to brainstorm the implications of our discoveries. There were 15 earth scientists at the site and not a single marine biologist, resulting in speculations on the sophomoric side.

"Certain species of anaerobic bacteria can metabolize hydrogen sulfide," John Edmond told us.

"You smell it all the time in swamp mud," I agreed.

We tried to envision the life-sustaining process at work. The sea water that seeped down through the fissures in the lava floor was under tremendous pressure—contacting the magma chambers, it became superheated and lost many normally suspended minerals while leaching out others from the basaltic rock—including sulfates—then vented back up to the seafloor carrying rich concentrations of dissolved hydrogen sulfide. This compound appeared as shimmering milky clouds as the hot suspension encountered the near freezing bottom water.

The abundance of hydrogen sulfide might explain the slimy mat cultures on the lava near the vents. Could these be paleo-microorganisms similar to those that had existed when the planet was young and the atmosphere was devoid of oxygen? And if so, did these bacteria form the basis of a food chain leading up to vastly more complex creatures such as clams and exotic fish? We needed some expert advice, so we radioed WHOI biologist Holger Jannasch. After we'd described all the animals encountered at Clambake I, Holger gave us careful directions. "First, take core samples in a grid pattern so we can analyze the organic qualities of the mud."

The four of us in the radio room shook our heads in frustration. Jack Corliss gripped the microphone. "No mud down there. It's all just bare lava, I'm afraid. Naked stone."

The speaker crackled with static a moment before Holger spoke again. "I don't see how that's possible. There must be some mistake."

Perched on a fissure, a sea anemone ekes out a meager living in the freezing waters of the Galápagos Rift outiside the nutrient-rich world of the hydrothermal vent.

We might just as well have described a healthy plant without leaves. But we had accurately depicted the virtually sterile desert of the sea bottom surrounding the bountiful islands of life we continued to encounter along the Rift. One site we named "Oyster Bed," which was dominated by chocolate-colored mussels. The "Dandelion Patch" looked like something out of a science-fiction movie: Its prevailing creature was a fleshy tube with a swollen head that seemed about to burst with seed. No one had a clue how to classify this oddity, and the marine biology books on board were no help whatsoever. The most intriguing site was the "Garden of Eden," where concentric rings of diverse organisms spread outward from the vents. The outermost ring was the domain of the fleshy dandelions and white crabs. Next inward, large white clams predominated. "Hedges" of tall, rather gruesome pale worms with blood-red maws and wriggling crimson heads clustered in the center of the vent, bathed in shimmering water.

As we continued to report our amazing discoveries by radio, I could sense the frustration in the voices of our biologist colleagues back on land. One of the true bonanzas of the marine life sciences was in the hands of geologists and geophysicists who hardly knew a lugworm from a sea urchin. But we did recognize that the sites teeming with unexpected life would forever alter the way science thought of the sea bottom. And we tried to speculate about the strange patterns that were emerging from our exploration.

MIT geophysicist Tanya Atwater presented a logical hypothesis one night as the gang enjoyed a sunset beer on *Knorr*'s fantail. The animals, she suggested, could have originated as free-swimming larvae and drifted down into the eternal night of the abyss. "Then they discovered the plumes of hot water, matured, and reproduced to found colonies."

Echoing a well-known precept of human heredity, in which a small group of pioneers pass their genes to a large population of descendants, Jack Corliss called the hypothesis the "Founder Principle." Whichever species reached the hydrothermal vent first—mussels, worms, clams, or the bizarre pinkish-orange dandelions—established the growing colony. Still, this did not explain a thriving multispecies site like the Garden of Eden.

There was a lot more theory out here than there was proof. One thing was certain, however: The terrain surrounding the colonies was devoid of sediment containing organic material produced by surface photosynthesis. Down on the Rift, all the complex life-forms depended on a food chain starting with primitive anaerobic bacteria that metabolized sulfides in absolute darkness. In turn, those bacteria supported a hierarchy of microorganisms that eventually produced more complex nutrients for the larger creatures. This was an ecosystem based on "chemosynthesis," and its implications were enormous.

There is a fundamental scientific principle that all major, complex life-forms on Earth depend on sunlight. Yet, we had just made startling discoveries that undercut that principle. Perhaps there were bodies in ours and other solar systems that might also harbor life far from the nurturing light of their parent stars. Planetary scientists speculated that one of Jupiter's moons, Europa, had a solid core and liquid sea beneath the frozen crust of its ocean. Twisted by immense gravitational tides, Europa was volcanic. Why couldn't the same processes that supported life down in the icy darkness of the Galápagos Rift also drive unknown biological processes on the volcanic floor of this Jovian moon? One evening before dinner, I was on deck with Jack Edmond. The tow cable angled down into the darkening water as *ANGUS* was tugged above the distant bottom, searching for more hydrothermal vents and their bizarre inhabitants. Jack grinned. "You know, Bob, it must have been like this sailing with Columbus."

I certainly felt that sense of discovery when I dove in *Alvin* to explore the living oases strung like pearls along the rugged central spine of the Rift. On March 10, 1977, I joined Jack Corliss on a dive. He was a bearded human grizzly bear—long, hair, more than six feet tall, and weighing about 240 pounds. A couple of the guys on the crew called him Jesus Christ, but did so behind his back. Somehow, Jack Corliss never took up more room in the cramped pressure sphere than an average-size person. But our pilot, Jack Donnelly, always admonished us to be careful about bumping switches when Corliss was diving.

Our first oasis appeared in the floodlights as the temperature-probe monitor pinged incessantly. A row of the orangish-pink dandelion creatures appeared, their swollen heads throbbing with filament coronas as they sensed *Alvin*'s pressure wave. This was a multispecies site, also densely colonized by large white clams and dull brown mussels. In the center of the colony, the warm water from the vents between the pillow lava shimmered past our viewports. "This is like flying through fountains," Corliss said.

The crew of the mother ship *Lulu* prepares *Alvin* for another dive on the hydrothermal vents of the Galápagos Rift. With amazing new discoveries coming almost daily, my colleague Jack Edmond commented, "You know, Bob, it must have been like this sailing with Columbus."

Jack Donnelly took a sample of the flaky material suspended in the clouds roiling from the vents, then rested the bottom skid on the lava so that I could examine the thick, mottled mats of organic material clinging to the pillows. I remembered pictures in undergraduate geology texts of calcified bacteria colonies in the world's earliest fossils discovered in the Australian desert. But these 20th century colonies of primitive microorganisms feeding on the harsh broth of hydrogen sulfide were alive, not fossils.

During our long descent to the bottom that morning, Jack Corliss had shared a nebulous theory engendered by our discoveries. Conventional scientific wisdom held that life had developed on Earth in the early planet's warm, shallow seas, which were rich in organic compounds containing carbon and sulfur. The energy and catalyst necessary to transform these compounds into self-replicating cells was thought to have been the intense ultra-violet solar radiation as well as the heat and electrical discharge of lightning. But Jack Corliss was beginning to speculate that life might have arisen at seafloor hydrothermal vents, the heat and chemical reactions providing the necessary ingredients.

As I gazed at the milky clouds of dissolved minerals swirling around the bacterial mats, I weighed Jack's hypothesis: What if life had arisen down here in the *deep* ocean, far from the influence of sunlight or lightning? What if the millions of years of hydrothermal circulation, cycling the mineral building blocks of life through undersea fissures, had actually initiated the miracle of evolution thought to have occurred in the shallow ancient seas?

Jack Corliss later became engrossed with this concept, widening his perspective at Oregon State University from geochemistry to fundamental and far-reaching themes. He energetically investigated the startling ramifications of his original shipboard musings, which eventually became a formal scientific theory. At a European conference in 1980, Jack and his Oregon State colleagues John Baross and Sarah Hoffman presented an elegantly reasoned paper: "An Hypothesis Concerning the Relationship Between Submarine Hot Springs and the Origin of Life on Earth." They presented tantalizing evidence that the earliest fossil record of cell-like structures appeared in Precambrian rock 3.8 billion years old, which geologists associated with undersea hydrothermal activity. The Oregon State team described how the constant cycling of water through hydrothermal vents in these ancient seafloors could have synthesized the complex organic molecules that formed the abiotic precursors to actual living cells.

I examine one of the giant red tube worms *Alvin* has recovered from a hydrothermal vent colony. Because these finds were so unexpected and we had no biologist on the expedition, we resorted to Tupperware containers filled with vodka to preserve our specimens.

I still find their supposition compelling. Certainly, hydrothermal vents combined the elements of energy, diverse mineral abundance, and ceaseless repetition necessary for the primal alchemy contemporary science recognizes must have occurred almost four billion years ago when life arose on Earth. Jack's hypothesis had other persuasive elements. The surface of the early planet was a dangerous place for life. Volcanism was almost constant. Meteor bombardment was much greater than in later times because the Earth had not yet swept its orbit clear of rocky asteroids, and the ancient atmosphere provided little protection from intense solar ultraviolet radiation. Any life that arose on the seafloor would have been shielded from the worst of these hazards. A lively scientific debate on this fascinating hypothesis still continues.

Such fundamental questions, however, did not concern us on this first expedition, during which we had virtually tripped over the living colonies at the hydrothermal vents. In fact, we were so unprepared for biological investigation that the lab had only one half-liter of formaldehyde. We sacrificed a couple of cases of duty-free Russian vodka *Knorr* had laid in transiting the Panama Canal to preserve our specimens in Tupperware, soup tureens, and roasting pans from the galley. Even though the specimen preservation was not ideal, the biologists at our home institutions were delighted by our discoveries. We had found several new species, including the big orange dandelion—an unknown type of siphonophore, a previously unknown, sedentary, distant relative of the free-swimming Portuguese man-of-war jellyfish.

Based on these discoveries, the National Science Foundation funded two follow-on expeditions to the Galápagos Rift so that leading marine biologists could exploit the rich research terrain we had pioneered. On these cruises, I was responsible for coordinating *Alvin* dives and conducting photo reconnaissance with *ANGUS*.

The biologists found that the nutrient-rich water around the vents formed a limited oasis, only 60 feet wide on average. The rare-steak flesh of the clams came from a concentration of hemoglobin higher than that of almost any other known creature, an accommodation to the oxygen-poor water bubbling from the vents. The clams' huge size was an evolutionary adaptation to the rich broth of nutrient those vents provided.

This expedition found that gigantism was a prevailing tendency among vent species. On a 1979 dive to a Rift site named the "Rose Garden," we encountered a thick stand of ghostly white tube worms with crimson maws swaying in the cloudy water. But these worms were much larger than the 18-inch species we'd first discovered. When one of the individual worms was taken up to the *Knorr*'s deck, it measured a full 12 feet long. The biologists almost danced with joy. This creature had no known relatives and came from an entirely new phylum; yet, it reproduced sexually, an indication that it had evolved over millions of years in the environment of the hydrothermal vents. Inside its semi-rigid tube, the worm played host to a newly discovered species of limpet filter-feeder, a small gastropod of a type only previously seen in ancient fossils—more evidence that chemosynthesis-based life had evolved over billions of years on the seafloor.

On the way to the site, I dipped into Darwin's *The Origin of Species* and thought of the young Englishman, miserably seasick aboard the small, spartan *Beagle* en route to the nearby Galápagos Islands. Those volcanic peaks had been shaped by a large "hotspot," a storm in the mantle forcing magma up through the overlying plate, as had occurred in the Hawaiian Islands and Yellowstone National Park. The discoveries Darwin had made on those islands had spurred his monumental theory of evolution. Now, 144 years later, American biologists who owed a vast debt to the English explorer were following his lead to explore an exciting aspect of evolution on the dark ocean floor nearby.

After the second 1979 cruise to the Galápagos Rift, a joint WHOI-Scripps expedition went to the East Pacific Rise at Latitude 21 North near Mexico's Baja California aboard the research ship *Melville*. We explored a fast-spreading tectonic fault, which my former French colleagues from the *Centre National pour L'Exploitation des Oceans* (CNEXO) had investigated the year before. We hoped to encounter more warm-water vents and associated colonies of life in order to prove the phenomenon was widespread. What we found was far stranger.

The initial *ANGUS* runs over this rift's central axis brought evidence of hydrothermal vents, and early photos revealed the typical clumps of white clams. On the first *Alvin* dives, my WHOI colleague Bruce Luyendyk and CNEXO's Jean Franceteau conducted a geophysical survey in which they saw weird pillar-like columns. Then French volcanologist Thierry Juteau and American geologist Bill Normark climbed aboard *Alvin* for a garden-variety gravimetric dive, but what their pilot, Dudley Foster, encountered on the bottom was anything but commonplace. The floodlights illuminated a bizarre forest of gnarled chimneys, most about six feet tall, belching thick black clouds of what looked like coal smoke.

"We've got a, ah...*locomotive* blasting out all this stuff," Dudley reported.

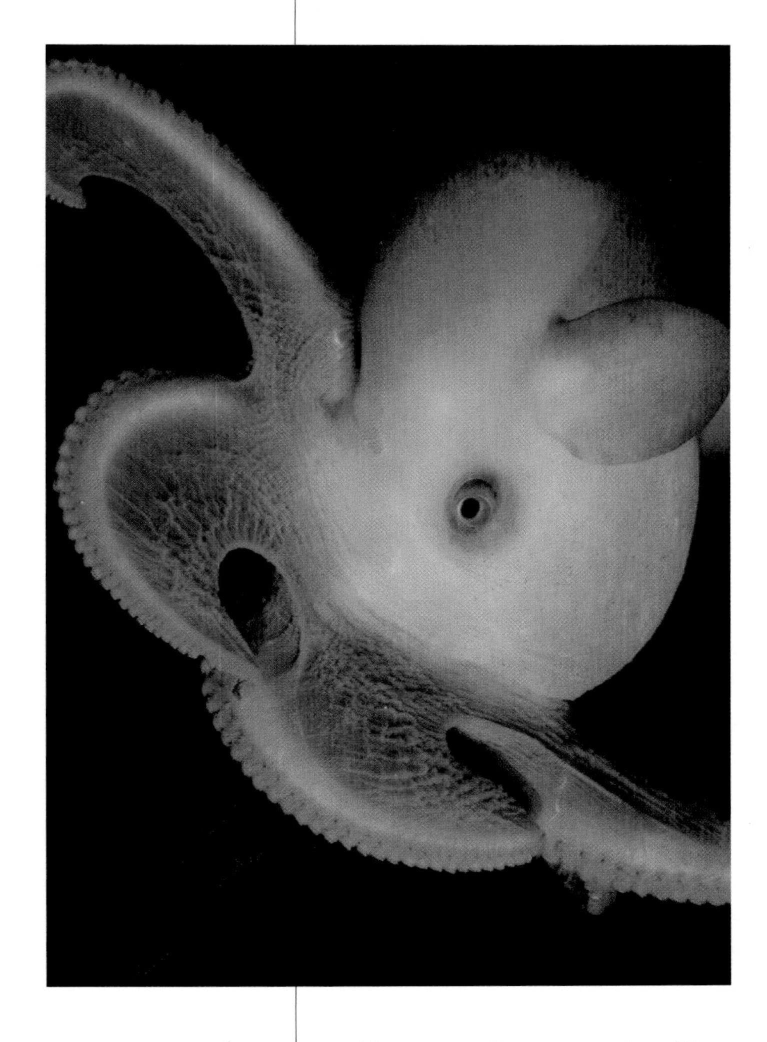

The surrounding warm water of the hydrothermal vents attracts deep-sea carnivores such as this purplish finned octopod, which uses its powerful arms and beak to seize and devour prey such as bottom-dwelling crustaceans and worms.

A "black smoker" superheated hydrothermal vent, photographed from *Alvin* on the 21 North seafloor chimney system, builds towers of almost pure crystalline zinc sulfide and creates a mineral-rich environment that supports variants of the chemosynthesis-based food chain we discovered on the Galápagos Rift.

Smoke, of course, does not form in a liquid. *Alvin* had found a previously unknown type of mineral suspension, much richer in concentration than the shimmering white thermal curtains we'd seen on the Galápagos Rift. These smoking chimneys were hot: *Alvin*'s temperature probe registered over 32°C (91°F) around the site. No one had ever heard of such warm water on the seafloor. That night when *Alvin* was hauled aboard *Lulu* for service, we were stunned to find the tip of the probe had melted off and the plastic wand was burned black. The crew repaired the probe and sheathed it in a heat-resistant material.

I was scheduled to dive the next day with Jean Franceteau.

"Watch yourself," Dudley told our pilot, Ralph Hollis. "That water is *hot* down there."

I was amazed to see the churning smokestacks the next morning. We were at the bottom of the Pacific, almost 9,000 feet down, but just outside *Alvin*'s viewports were billowing cylinders reminiscent of mud-and-wattle chimneys in undeveloped villages.

Jean gazed silently, then spoke: "They seem connected to hell itself."

Minding Dudley's warning, Ralph was cautious as we edged toward the first belching smokestack.

"Let's just get a reading before we go much closer," he said.

"Roger that," I said, licking dry lips. We didn't want sudden heat damage down here.

Ralph's wariness was not unwarranted. Our newly armored temperature probe gave a reading of 350°C (662°F) at the snout of the billowing black geyser, higher than the melting point of lead. Only ten feet away from *Alvin* was a fountain of superheated water that would have instantly melted our viewports on contact. Had that happened, there would not have been time to call Mayday. The people topside would have never learned how we died.

We dove on the chimney vents for the next two weeks, identifying two basic types, "black smokers" and "white smokers." The really big ones stood 30 feet above the surrounding lava, making it vital that we conduct a careful sonar and visual approach to each site. Analyzing the material of the chimneys, the geochemists found almost pure crystalline zinc sulfide.

These smokers belched out incredible volumes of sulfide-rich minerals because the magma was so close to the sea bottom. And the 21 North chimney ecosystems differed from those of the Galápagos Rift. Although we found the expected giant clams and tube worms, we also discovered unique evolutionary variants of more common fish and snails that had adapted to the hot, mineral-rich environment over millions of years. Our expeditions opened the door to a new and unexpected branch of evolutionary study. Years later, the British marine scientists John D. Gage and Paul A. Tyler noted that "deep-sea scientists were astounded by the unexpected" nature of our discoveries in the Pacific. Science had traditionally held that the oceans, which comprise almost three-quarters of the planet's surface, nurture a bounty of wondrously complex life. But it was only after our exploration in the eastern Pacific that we began to understand how prolific the life-sustaining oceans actually are.

Even though I now conduct most of my ocean "diving" robotically with ROVs, I have very fond memories of the hundreds of voyages to the sea bottom I was privileged to make aboard submersibles. A deep dive, especially in nutrient-dense waters, always provided a spectacular encounter with the open ocean's "water column," a vertical slice through Earth's most abundant biosphere.

When *Alvin* slipped free of *Lulu* preparing to dive, I sometimes saw silvery flying fish flash by and disappear into the air above, then extend their pectoral fins like birds' wings to sail away. This marvelous adaptation allowed them to escape predators such as the big striped mackerel and even larger yellow-fin tuna that sometimes curiously nosed close to our viewports. Descending toward the bottom at a hundred feet per minute, *Alvin* encountered creatures few land dwellers have ever seen.

Initially, the water retained its sunny blue. But the world of warmth and light steadily receded. By the time the sea had darkened to indigo, then to featureless black, our external lights revealed the deep scattering layer. This ocean-wide environment hosted a myriad of tiny creatures contained within their own ecological system. Larval fish, crustaceans, mollusks, and gastropods swarmed in incredible profusion, feeding on photoplankton and protozoa. In turn, shrimp, squid, jellyfish, and fish preyed on the smaller species. This teeming zone of life acted like one vast, colonial organism, collectively rising to the surface to feed at night and sinking back to about 2,000 feet during the day to escape larger predators. The layer easily reflected or "scattered" the pulses of early depth sounders, producing false readings.

Next, one entered a blizzard of marine snow, the organic detritus ceaselessly falling through floodlights from the water above. This material was composed of fecal pellets, dead plankton, and the decomposed remains of creatures from the deep scattering layer. The snow might fall for months, before delivering its precious cargo of organic compounds to the nutrient-poor bottom sediments.

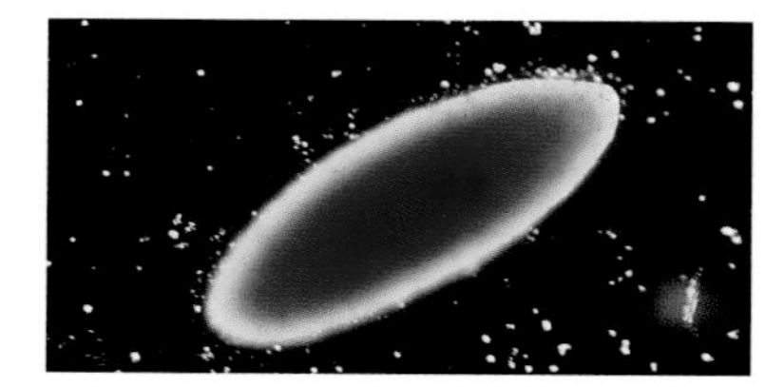

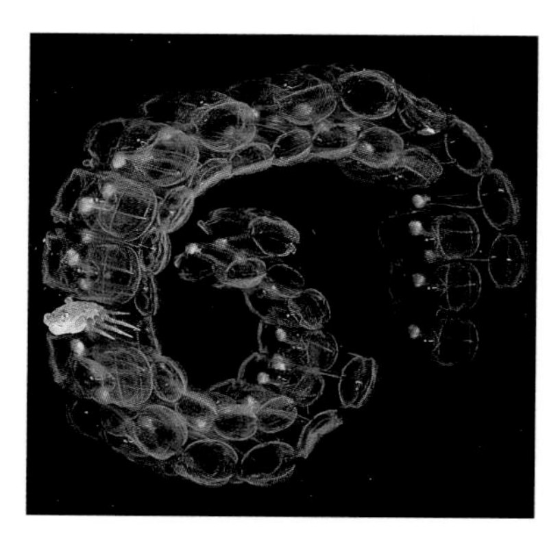

"Marine snow," here a fecal pellet of a microscopic drifting creature magnified 250 times glowing in polarized light, and the remains of larger organisms drop ceaselessly from the surface to the seafloor, eventually building deep nutrient-rich sediment layers.

A sargassum crab rides the Gulf Stream current on a translucent salp chain.

Passing through the scattering layer, I always preferred a dark descent with the external floodlights extinguished. We usually encountered twinkling constellations of lantern fish. Sometimes they numbered millions in a single school, flashing in all directions at our presence. There was little a Hollywood special effects wizard could do to top this display of cold green fire. We'd also glimpse bioluminescent viper fish, miniature monsters only inches long, their gaping jaws lined with saber-sharp fangs, stalking even smaller fish.

Then we'd drop through the barren oxygen-minimum zone where few bioluminescent animals existed. This lonely void extended to about 5,000 feet, where the ambient pressure reached almost 200 atmospheres. In this unlikely environment, we'd sometimes get a random sonar ping, possibly from a giant squid or blind gulper eel swimming nearby. In any event, the pilot would have already killed the dim red light inside the pressure sphere when we reached 1,200 feet so that we would be able to identify the source of the sonar activity, or it us.

This was prudent. On a dive off the Bahamas in 1967, a swordfish had flashed out of the darkness to attack *Alvin*. The creature drove its bony sword deep into the submersible's fiberglass hull. By sheer chance, no critical plumbing pipes or electrical cables were severed. Pilot Marvin McCamis dumped ballast and performed an emergency ascent. On the surface, the huge fish, its sword still skewering the sub, thrashed in agony from the sudden pressure change. During these death throes, the fish snapped off its impaled sword. But the recovery scuba divers swimming warily around *Alvin* looped a line around the dying fish's tail. The rich aroma of fresh swordfish steaks filled *Lulu*'s little mess hall that night.

I certainly never wanted to tangle with a giant squid at any depth. These huge creatures were known to frequent the lightless void below the deep scattering layer. In the cold, nutrient-rich water of the Humboldt Current in the East Pacific off South America, the huge squid *Dosidicus gigas* normally grew to lengths of 12 feet and weighed 350 pounds. Its thick grasping tentacles were lined with sharp-edged circular suckers, which often left rows of nasty scars on sperm whales. But that species of squid was not the ultimate sea monster. This distinction went to the true giant squid, *Architeuthis princeps*, which reached lengths of 55 feet in the western Pacific. I've often wondered whether a squid that size could tear apart a small submersible like *Alvin* in the relentless grasp of its powerful tentacles. Jules Verne certainly speculated that a monster squid could give Captain Nemo's *Nautilus* a hard time.

At depths of 9,000 feet, we sometimes encountered solitary anglerfish. Their appearance was always bizarre: At least half the fish was head, dominated by a fearsome cavern-like mouth ringed with razor-sharp teeth. A weird flexible rod-like organ arched from the fish's forehead, tipped by a pulsing green bioluminescent lure. They might cruise through the silent darkness for weeks before attracting a hapless abyssal rattail fish to their bait. Unlike the tiny viperfish higher up the water column, these creatures could reach lengths of four feet. When one that size drifted by the viewport and peered inside as if sizing up my face for lunch, I always felt a chill. Sighting them reminded me that the deep sea was a true wilderness, which humans had hardly begun to explore.

Except on the freshly extruded lava of tectonic rifts, *Alvin* always kicked up billowing clouds of sediment on reaching the bottom. I have seen this same featureless expanse of marine snow on the seafloors of all the world's major oceans. Falling ceaselessly from the bountiful upper layers, the organic material has accumulated about an inch a millennium over millions of years. If extraterrestrial visitors ever probed Earth with instruments that made the oceans invisible, they would view a planet whose surface was almost three-quarters gray mud.

Bottom creatures were always scarce, except around hydrothermal vents, but I usually saw a few living animals or their tracks. Holothurians varied in shape and size from individuals that looked like purplish sections of radiator hose, to elongated flapjacks with crenulated edges. One of the most interesting was the elasipodid, which, unlike its crawling cousins, scrunched up its body into a spring and popped into the current to hover briefly before descending once more to the sediment. The larger of this species had a double row of ventral tube feet and a protective spiny back, and trudged purposefully along the bottom as if late for an appointment.

I was always surprised to find fish in an environment where food was so scarce, but I generally caught a glimpse of a scavenging eel-like macrourid undulating along the muddy bottom, or a morid fish probing the oozy sediment with its sensitive pelvic spines. Depending on the location, there were deep-sea sponges, slow-growing corals adapted to the frigid darkness, or small clusters of sea anemones. The diversity of bottom fauna depended in large measure on the amount and quality of marine snow falling from the surface. On the seafloors beneath surface layers thick with life, the sediment was denser in nutrient, and bottom dwellers were more

Tiny Atlantic lanternfish school in the millions, rising at night to feed on plankton. These and other creatures form the "scattering layer," so dense it disrupts sonar.

Giant clams and a single red shrimp colonize vents at 9° north latitude. When the volcanic upwelling fueling the vents subsides, animals starve in the icy darkness.

abundant. But in areas of the open ocean far from continental rivers or nutrient-rich currents, the bottom was often a near desert, as we found on the seafloor of the open Pacific where *Yorktown* rests.

On virtually every dive, I sighted a little red shrimp creeping across the bottom. I've seen one in the wreck-strewn Iron Bottom Sound off Guadalcanal, near the *Titanic*'s grave, and on the floor of the Mediterranean. When diving with someone for the first time, I'd joke that *Alvin* had a hidden compartment from which we released the shrimp. "We trained him to get back in before we surface," I'd manage with a straight face.

Each year, countless air-breathing dolphins and sea turtles drown as factory ships scour the world's oceans for tuna and squid, such as this victim caught in a tuna-fishing net.

But there is a serious aspect to the comic shrimp's walk-on role. Shrimp are complex animals compared to corals, sponges, and eyeless holothurians. If a shrimp could make a living down among the dregs and crumbs of the upper ocean, this might mean the sea above is still relatively healthy. Maybe bottom-dwelling shrimp, similar crustaceans, abyssal fish, and other higher species are like canaries in the coal mine. They are among the first species to disappear when seafloors become gravely polluted with sewage or the toxic effluent of our industrial society.

I think that one of the great tragedies of the last millennium is that—like air and freshwater pollution on land—the defiling of the Earth's oceans proceeded unchecked for hundreds of years before the public recognized the problem in the 1960s. By then, terrible damage had been done. One need only dive to the dead floor of the Adriatic or in the once pristine waters off New Caledonia in the Pacific (now a cauldron of toxic heavy metals from mining spills) to find irreversible pollution. But we can't point our finger at foreigners. Americans have been just as wanton in their treatment of the sea, which was considered a bottomless sewer since the first English colonies were established. When I was a member of the Boston Sea Rovers scuba club in the 1960s, we sometimes dived through raw sewage, trying to avoid the worst of it, as if the stinking tide were a natural problem to be overcome.

Only after the creation of the Environmental Protection Agency in 1970 did some major coastal cities stop pumping their untreated sewage directly into the ocean. But paper mills, chemical plants, oil refineries, and other industries routinely discharged toxic waste into nearby rivers, which followed their ancient courses into the sea where the poison spread across the breeding grounds and nurseries of birds, fish, and marine mammals. Not until "top predator" species high on the chain of predation, such as tuna and swordfish, became contaminated with chemicals like PCB (polychlorinated biphenyls) did we begin to face the gravity of this problem. They accumulate persistent, fat-soluble organic compounds from the fish they feed on and can pass this dangerous concentration on to us when we eat their flesh.

Such environmental disasters should be alarm bells alerting us to an obvious, but previously unrecognized, truth: Eventually, every toxin or pollutant we loose in the biosphere makes its way into the ocean. Insecticide sprayed on a Pennsylvania pear orchard drains into the Chesapeake Bay where it is ingested by grass shrimp, who are eaten by perch, who fall prey to striped bass, who are hunted offshore by yellow fin tuna on their winter migration south. Similar patterns prevail around the world. The sea and its creatures cannot go on absorbing our toxic wastes indefinitely.

But even as the modern environmental movement gathered momentum, serious threats to the ocean ecosystem continued. Rampant overfishing, especially by "dragger" trawlers larger than some World War II cargo vessels, has decimated stocks throughout the world. When Leif Eriksson explored Vinland a thousand years ago, his men could wade from shore and scoop up fat salmon in their shields. Today the once thriving Grand Banks fishery of Atlantic Canada is fighting for its life. The New England fishing industry that supported tens of thousands of families until the 1950s is also moribund. Diving in *Alvin*, I have frequently seen the savage scars bottom trawlers cut across the ocean bottom, destroying a delicately balanced ecosystem that will take centuries to heal. This pattern is repeated in other nations. Too many boats are hunting too few fish, using illogically wasteful techniques.

Only recently did shrimpers and offshore commercial fishermen give in to mounting environmentalist pressure to adopt less destructive methods. Many American shrimpers now fit their nets with "excluder" devices to allow air-breathing sea turtles an escape route when caught. Most of the canned tuna on American supermarket shelves proclaims that no dolphins died in taking the catch.

Although environmental awareness has flourished in the past 30 years, we still have a long way to go. Several Asian fishing fleets still follow disastrous practices that will drive Pacific dolphins and sea turtles to extinction if not changed. These fleets set drifting gill nets at night, some monofilament curtains 30 feet deep and 40 miles in length. The "target" species are tuna and squid. But each morning when the miles of yellow net are winched aboard the factory ships, dead dolphins are gaffed aside and tossed overboard. This same tragic pattern occurs for sea turtles when drift nets are employed in warm waters.

I don't see any reason the international community should tolerate such slaughter on an industrial scale that has not been seen since the days of wide-open commercial whaling, which ended in the 1950s. While some environmental activists doubt that the drift-net problem will ever be solved, I prefer a more positive approach. Once the public became aware of unchecked commercial whaling's devastating impact, international conventions virtually shut that industry down. Today, whale species that were near extinction are on the rebound. The lesson here is that the living ocean has the capacity to heal itself if not pushed beyond ultimate limits.

But drift nets are not the only manmade hazards marine mammals and reptiles face. Ubiquitous plastic—in the form of garbage bags, foil-coated party balloons, and six-pack ring connectors—kills countless thousands of seals, dolphins, manatees, and sea turtles each year. Fur and harbor seals are powerful divers, especially prone to strangulation when swimming through floating nylon net fishermen thoughtlessly discard. Baby sea turtles are voracious surface feeders. Floating crumbs of otherwise harmless Styrofoam make a deadly meal for these harmless, endangered creatures.

Coastal ecosystems are especially vulnerable to oil pollution. Although the world has not forgotten the oil spill from the Exxon *Valdez* that devastated Alaska's

***Following pages:* Anthias fish swim through green tree coral beneath the hull of an outrigger canoe crossing Goodenough Bay in Papua, New Guinea. Pristine coral reefs are endangered, with threats ranging from undeveloped nation's fishermen using poison to cruise liners dropping anchor.**

Prince William Sound in 1989, I guess that global dependence on petroleum has led to an almost blasé attitude toward the environmental hazards huge tankers still pose. It's easier not to think of the risks that transporting almost a billion barrels of crude oil each year entails. When Valdez became the Alaska pipeline oil terminal in 1977, the Alyeska Pipeline Service Company told state and federal officials the tankers using the port would be double-hulled, a sound precaution against oil spills. But the Exxon *Valdez* and almost all the other immense tankers taking on oil there were single-hulled. They were ecological time bombs.

The tanker Exxon *Valdez*, carrying 1,264,155 barrels of Alaskan oil, struck Bligh Reef at midnight of March 24, 1989, the result of multiple human errors by the crew. The hull was ruptured so badly that crewmen saw a black gout of crude oil spewing 50 feet into the chill night air from one of the tanks. Within 20 minutes, more than 100,000 barrels of oil had hemorrhaged into the sound. Over the next hour, Captain Joseph Hazelwood ground the tanker back and forth over the rocky reef

trying to free his vessel. Before ships could be marshaled to offload the remaining cargo, the huge tanker had spilled 10.8 million gallons (almost 260,000 barrels), the worst pollution disaster in maritime history.

Prince William Sound was ravaged. Viscous oil drifted widely across the deeply indented coastline, fouling beaches and coves and settling into the herring, salmon, and halibut spawning grounds on which the local fisheries depended. For weeks, television news reports showed workers in yellow coveralls struggling to save dying oil-poisoned sea otters and birds. During the long days of the Arctic summer, those same workers tried vainly to sop up the oil with rags and straw. It will be decades before the Sound's ecosystem recovers. In 1994, Exxon was heavily fined and ordered to pay compensation to over 40,000 Alaskan commercial fishermen and Native Americans whose livelihood and way of life had suffered.

Green sea turtles mate in the limpid waters inside the coral reef of Sipadan Island near Borneo. Reefs provide the invaluable mating grounds and nurseries for thousands of species in the tropic oceans worldwide.

But I think people have forgotten that single-hulled tankers continue to ply the world's busy sea-lanes. Every year, it seems, a tanker runs aground, spilling its toxic cargo. I am hopeful, however, that international public opinion will soon reach such a point of indignation that this needless tragedy will no longer be tolerated. After all, with the leadership of American and European environmentalists, seemingly intractable problems of city pollution caused by leaded gasoline and inefficient car engines, air pollution, and toxic ground water sites were resolutely confronted for the first time in the 1970s. And only 30 years later, many of these difficult issues have been resolved.

Unfortunately, we probably don't have 30 years to save the world's threatened coral reefs. These delicate ecosystems are as biologically complex as rain forests, but have been much less widely studied. Because they are under water, coral reefs are literally out of sight, and I suppose we have long tended to ignore the dangers they face.

A coral reef is a vast colony of living creatures, an aggregation of individual marine invertebrates called polyps. They each form a protective rocky skeleton of calcium carbonate and are anchored to the lifeless skeletons of innumerable previous generations beneath them. Coral reefs are restricted to shallow, warm (but not too warm) water, generally within the tropics. Most living coral will not tolerate seawater temperatures lower than 20°C (68°F).

Coral polyps host symbiotic algae called zooxanthellae with which they exchange nutrients in an efficient, mutually beneficial manner. Individual polyps feed at night, opening their tiny maws to trap free-swimming zooplankton with minuscule tentacles. Because reef coral cannot thrive without sunlight to fuel the algae's photosynthesis, the living polyps must be near the surface.

A coral reef, whether fringing a coast close to shore, an offshore barrier, or the ring protecting an atoll's shallow lagoon, provides a habitat for a stunning diversity of life. Algae accumulating outside the polyps support herbivorous fish, the flamboyantly striped, spotted, and gaudily splotched species we associate with tropical waters. But reefs also host almost countless species of algae-feeding invertebrates, mollusks, bivalves, sea urchins, and starfish, to name a few. Further up the food chain, crabs, moray eels, spiny fish such as wrasses, groupers, and sharks inhabit reefs.

Reefs play another vital role in the interconnected ocean ecosystem. Many species of fish congregate along reefs during their annual spawning season. The larvae that emerge from the fertilized eggs are then swept out to sea through channels in the reef. Without the structure of the reef, these species would be unable to propagate.

For all their rock-solid appearance, coral reefs actually present a fragile veneer of living tissue. The myriad individual polyps and their symbiotic algae are extremely vulnerable to the pollution and physical battering that assaulted reefs throughout most of the 20th century. Coral polyps quickly die when coated with the river-borne sediment resulting from the slash-and-burn agriculture practiced so widely in the developing nations. Untreated sewage also kills coral. Obviously, the destructive "mining" of coral reefs for sand and rock to make concrete leaves huge dead gaps in the once living colonial structure.

Cruise ships anchored on coral reefs for decades so that their snorkeling passengers could enjoy the splendor beneath the surface. But the ships' huge anchors, as well as those of smaller yachts, gouged the living reef coral of the Caribbean and Gulf of Mexico for years before mariners became aware of the harm they were causing. The widespread use of poison among developing nations' fishermen has also proved especially destructive to coral reefs.

During the International Coral Reef Symposium in Guam in 1992, scientists presented a grim picture. At least ten percent of the world's coral reefs had degraded beyond recovery, mainly as a result of human activities. Perhaps a third more, especially in Asian-Pacific waters, are threatened with extinction. Another 30 percent are likely to decline within the next 20 years, including reefs in the Caribbean, along the Florida Keys, and the east coast of Africa.

This was a global alarm bell, which spurred the development of an environmental effort to save our planet's reefs. Small island nations that directly depend on coral reefs for fisheries and tourism met in Barbados in 1994 to form the International Coral Reef Initiative. Their goal is to reverse the most destructive sewage-disposal, agricultural, and fishing patterns and undertake public education on the problem.

That same year, as part of annual JASON Project, I assembled a small team of scientists and students to introduce a much larger audience of students to the menace facing our coral reefs. It will be today's youth who will have to complete the long and difficult task of conservation needed to restore the world's coral reefs to health.

On March 7, 1994, reef specialist Dr. Jerry Wellington stepped off the back of a dive boat drifting through limpid blue water, 40 feet above the living coral of South Water Caye in the Caribbean republic of Belize. Jerry wore a bulbous Plexiglas LAMA Bubble helmet that allowed him all-around vision and permitted him to speak normally. The helmet was equipped with a microphone and he carried a color video camera. For the next hour, Jerry took several hundred thousand North American students on a field trip along the reef. His video and audio feeds were relayed from the boat to a shore station, then uplinked to a communication satellite and

Students in the JASON Project annually "go" on expedition, most by the telepresence of satellite video. Some can maneuver our ROVs from workstations at schools and museums near their homes. One goal of the project is to train a new generation of scientific explorers.

downlinked to a score of JASON Project sites where the students and their teachers had gathered. Dr. Wellington's lesson went beyond the gee-whiz diversion too often found in American science classes. The students and teachers who had qualified for the Project had already studied the diversity of species on tropical reefs. They were beginning to understand practically a simple but monumental concept gleaned from any decent science text: Nothing exists in isolation within Earth's incredible biosphere.

Today's class was complex. Many reefs around the world were suffering from coral bleaching, in which the polyps' symbiotic algae were expelled, leaving the coral organism weakened and devoid of color. Two main hypotheses addressed the causes of this problem. One blamed rising seawater temperatures, perhaps associated with long-term global warming. The other theory connected ultraviolet radiation—possibly related to atmospheric ozone depletion—to the loss of coral algae. This week's study would focus on those two theories.

Digging into such complex material was a real challenge for grade school children; I was proud that I could help bring such provocative material to these youngsters. I wanted them to actually struggle with such issues because I was confident that somewhere out in that audience seated before the big video projector screens there were boys and girls who would never forget these lessons. A few would devote themselves to the difficult, often frustrating vocation of scientific exploration. But many more would become dedicated to the goal of leaving their own children a healthier planet than they had inherited from us.

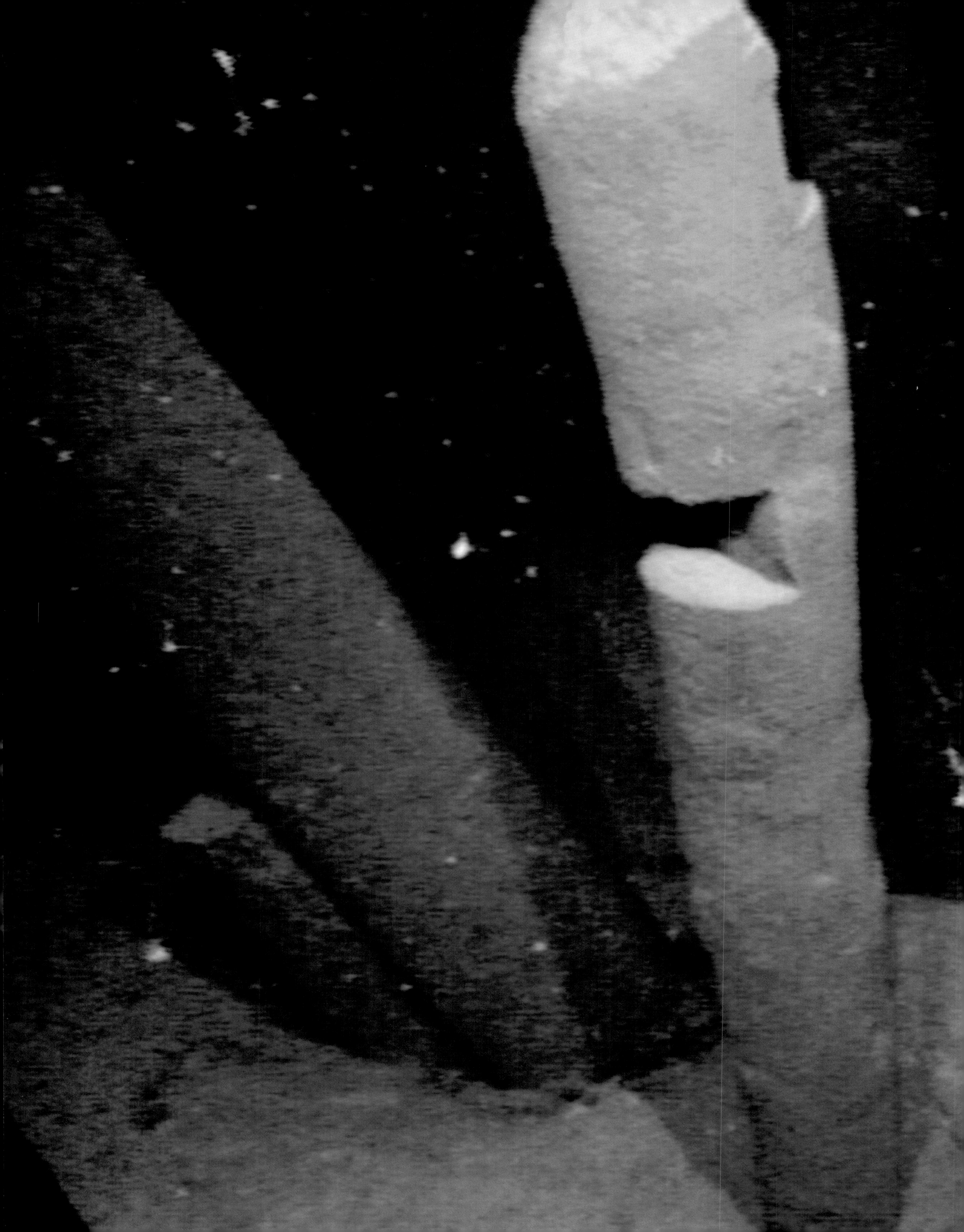

IX The Museum of the Deep

Preceding pages: **A pristine wooden stanchion with a hand-carved notch rises beside the standing mast of the perfectly preserved sunken ship that has rested 1,500 years in the deep waters of the Black Sea. The ship's discovery confirmed the theory that these sterile, oxygen-depleted depths protect ancient shipwrecks from wood-boring organisms.**

A Turkish shipwright works his art on the frames of a trawler being built near Sinop. In ancient times, the process was reversed, hull planking joined with mortises and tenons coming first, and later strengthened by internal framing.

I went to bed at my Istanbul hotel around midnight on August 30, 2000, exhausted from the last hectic weeks of pre-expedition planning. But I was just too excited about finally starting this cruise with a proper research ship—*Northern Horizon*, a full crew of scientists, engineers, and technicians, and a stable of state-of-the-art sonar and imaging vehicles to fall asleep. So I reviewed the months leading up to my arrival in Turkey.

Our 1999 expedition had confirmed geologists William Ryan and Walter Pitman's theory that the Black Sea had been a freshwater lake, which had been inundated about 7,500 years ago by a catastrophic flood. One of the goals of this year's expedition was to search for evidence of human habitation on the flooded former lakeshore along the narrow fan of shallow submerged coastal plain west of Sinop.

To meet a second goal we would head farther offshore. Ever since I had read Willard Bascom's insightful book *Deep Water, Ancient Ships*, I'd been obsessed with exploring the Black Sea's depths, hunting for well-preserved shipwrecks from antiquity. Bascom, an oceanographer at the Scripts Institute of Oceanography in San Diego, was the pioneering scientist who linked the Black Sea's unique hydrology with the Sea's unparalleled potential as an archaeological museum.

Often compared to a bathtub without a drain, the Black Sea's deep basin had filled with dense saltwater thousands of years ago, then was overflowed by the European rivers to the north. Over the millennia, a surface layer of lightly saline water formed down to a depth of about a hundred meters (328 feet). Sealed from above by fresher water and trapped by the narrow strait of the Bosporus, the immense volume of the sea's salt water, unable to circulate, became anoxic, or completely depleted of dissolved oxygen, and unable to support life, including the wood-boring worms that relentlessly consume every plank, mast, and boom of shipwrecks exposed to oxygenated water.

As I read and reread Bascom's fascinating book, a dream had formed. Ships had sailed between present day Turkey and the Crimean Peninsula since the Bronze Age. We would concentrate our search for wrecks along these ancient shipping lanes, down in the anoxic depths that could reach over 2,000 meters (6,562 feet). When I really allowed myself to dream, I visualized discovering a perfectly preserved Homeric ship such as Bascom evokes in his book: "Somewhere, far out beneath the wine-dark sea of Ulysses," he wrote, "there lies an ancient wooden ship. It sits upright on the bottom, lightly covered with the sea dust of twenty-five hundred years.... The stub of the mast still remains..."

Our expedition shuttle boat leaves the Turkish Black Sea port of Sinop, heading for the research vessel *Northern Horizon*. After weeks of frustration and bad weather, we discovered four ancient wrecks, including the perfectly intact vessel in the anoxic depths.

Because the workhorse ROV *Jason* increasingly had to be shared with other scientists, I decided to create a new set of vehicles for the Institute for Exploration's exclusive use. Underwater archaeology requires precise three-dimensional imaging of sites, using a platform equipped with color video and electronic still cameras that produce the digital mosaic photomaps academic experts need to assess the find. For close-in visual inspections and artifact retrieval, we needed a small, nimble remotely-operated vehicle, a younger cousin ROV to *Jason*. I hired robotics wizard Jim Newman, and in spring 2000, we began developing these advanced imaging systems. The first, *Argus*, is an optical tow sled, which is connected to the surface ship by a fiber-optic cable. Housed in a strong stainless-steel rectangular cage, the vehicle has three pivoting video cameras, floodlights, an electronic still camera, thrusters to hold position above a bottom site, and a scanning sonar. *Argus* is designed to be the "parent" to the mid-size ROV *Little Hercules*, to which it can be connected by a short fiber-optic tether. Fully maneuverable from its multiple thrusters, equipped with a broadcast-quality video camera and sophisticated lights, "*Little Herc*" can be fitted with a remotely operated manipulator arm to retrieve artifacts. We also planned to develop our own deep-towed side-scan sonar, ECHO, but would use the Woods Hole *DSL-120* until the new system was available.

I was confident that the technical and academic team we had assembled for the expedition was up to the challenge. Jim Newman, Dave Wright, and Charlie

Smith were first-rate theoretical and practical engineers who could handle any of the inevitable equipment glitches we'd encounter. We had a proven squad of *Little Hercules* pilots in Martin Bowen, Philip "PJ" Bernard, and Craig Elder. And *Argus* would be in loving hands with Tito Collasius, Jim Newman, and Dave Wright standing watch. Dr. Fred Hiebert of the University of Pennsylvania would be the lead archaeologist on the search for Paleolithic sites, while nautical archaeologist Dr. Cheryl Ward of Florida State University would be the expert on ancient shipwrecks. Above all, I would depend on my IFE deputy Dwight Coleman, and my long-time chief-of-staff Cathy Offinger.

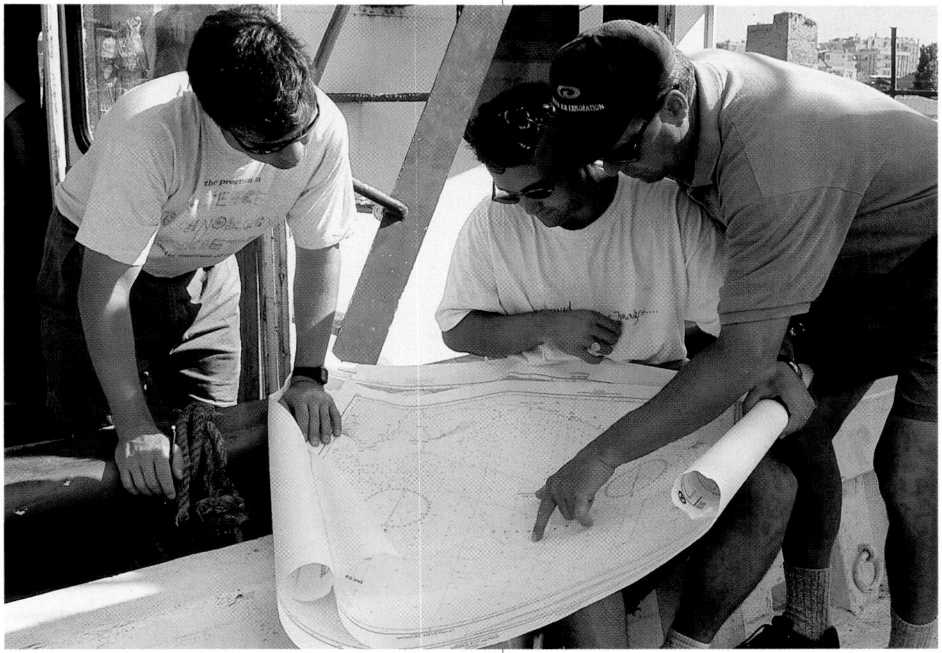

I had again managed to tap the generous funding of the National Geographic Society, the National Oceanic and Atmospheric Administration, the Office of Naval Research, and the J.M. Kaplan Fund. I had already decided that this cruise would be an extensive reconnaissance effort that would not include a request to the Turkish government to recover artifacts. I first wanted to prove the feasibility of deep-water archaeology by capturing images of any submerged Paleolithic sites or ancient shipwrecks before attempting physical contact. I've always believed the seafloor is and should remain a museum and that only a minimum of artifacts should be recovered for dating purposes. If we did encounter a well-preserved wreck, of course, I would ask permission to take a wood sample for radiocarbon dating.

I confer with my deputy Dwight Coleman (center) and MIT graduate student Brendan Foley (left), consulting a detailed bathymetric chart of Turkey's Black Sea coast. We were searching for traces of riverbeds submerged in the Great Flood 7,500 years ago.

As the small boat carrying the National Geographic film crew and me approached *Northern Horizon* riding at anchor the next morning, the ship looked good with her hull freshly painted bright red and white. I noted the big blue square-frame derrick on the stern for launching and retrieving *Argus*, and the smaller hydraulic cherry-picker crane folded on the portside fantail we'd use for *Little Herc* and the *DSL-120* sonar. This well-equipped expedition ship was a far cry from the fleet of little fishing trawlers we'd worked with last year out of Sinop.

After our slow passage north through the Bosporus we increased speed, and turned east toward our first survey site about 40 nautical miles away. That afternoon, I called the expedition crew together in the mess hall to brief them on our search strategy.

"We've got to look in the right places for the correct clues to possible Stone Age habitations," I told the team. "By 'right places,' I mean livable areas as they were 7,500 years ago, particularly river valleys flowing into the ancient lake." Our study of bathymetric charts had shown that most of the ancient coastline was nearly sheer cliff from the present-day highlands to the depths. There were only two narrow fans of gently sloping land between the Bosporus and Sinop that had probably been bisected by streams and rivers, which were now submerged beneath

Martin Bowen (center) troubleshoots a problem with our ROV *Little Hercules*, while Mark DeRoche looks on with concern. The glitch, which we first thought was a show stopper, was simply a broken fiber-optic cable connection. We repaired the break and were back exploring within hours.

up to a hundred meters (325 feet) of water.

"That's where we'll search," I said.

Our tactic would be to run long, slow *DSL-120* sonar lines from west to east, and back again, hunting for submerged riverbeds. Then we would narrow our sonar tracks along these ancient waterways, searching for any unusual angular formations suggestive of foundation stones, such as Fred Hiebert had found in earlier land surveys near Sinop. Once promising "targets" had been identified, we would switch to visual mode, first using *Argus* on a video survey of the site, and then closing in with *Little Herc*.

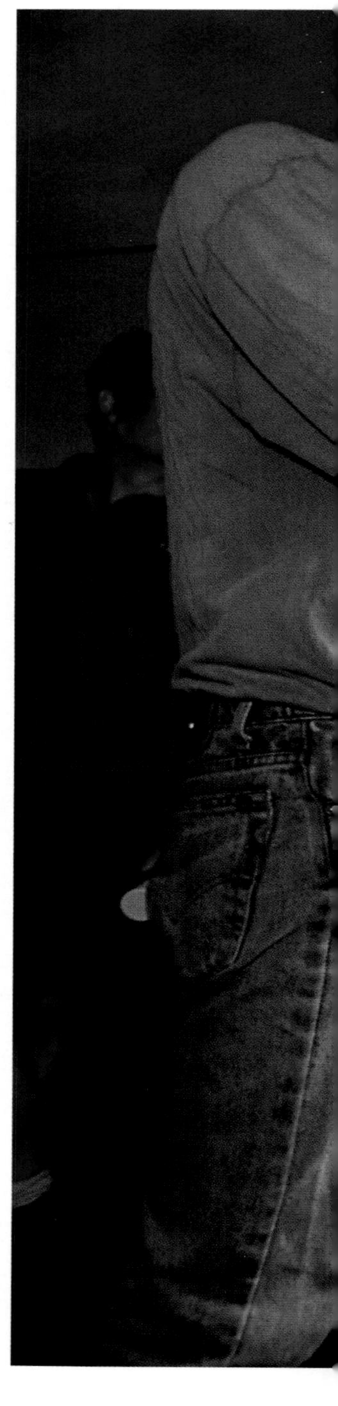

Our deep-water shipwreck survey would employ the same pattern: *DSL-120* sweeps followed by visual inspection using Argus and Little Hercules on targets. By mid-morning on September 2, we were ready to run our first long west-to-east track, working into shallower water. Although it was a splendid calm summer day with a sparkling blue sea and fluffy clouds, I spent long hours in the dim control room on the ship's main deck, watching the familiar charcoal-gray bands of the side-scan sonar scroll down the monitor. The DSL crept below us at a speed of only 2.5 knots, climbing slowly to a depth of 95 meters (308 feet), revealing a lineated bottom of what appeared to be sand waves, which led to a flat east-west mesa of cap rock edged with seaward-facing bluffs.

We continued our long sweeps, hunting unsuccessfully for submerged streambeds. But we were finding some intriguing targets. "Oh," I called, pointing at the monitor. "Upper right. There's a bell ringer." The watch team in the control room all turned from the navigation station, sonar, winch control, and data-logging station to note the small gray shape sliding down the monitor like a cluster of tiny grapes. We'd seen those before in the Mediterranean: heaps of amphorae, evidence of a possible ancient wreck. The hours passed. More possible grape-cluster shipwrecks appeared, but I was disappointed that we hadn't found any obvious stream channels or rivers flowing into the ancient lake at a depth of 155 meters (503 feet), which was where we would have expected to find them. Late that night our patience paid off, however, when the sonar beam intercepted an obviously meandering river channel on a flat plain in our survey area just east of the small port of Turkeli. Now our goal was to map as much of the river channel as possible from the ancient shore to the modern coastline, running sonar lines perpendicular to it as we searched for structural targets on either side. We found a number of intriguing targets, including one with rectangular blocks about five meters (16.4 feet) along one side.

At noon on September 3, we completed this sonar survey and I decided to launch Argus. "We've got a good shot at a real hard target," I told the crew. It lay where two river channels converged, an ideal site for a human habitation. We were switching over to visual mode earlier than planned in order to exploit this

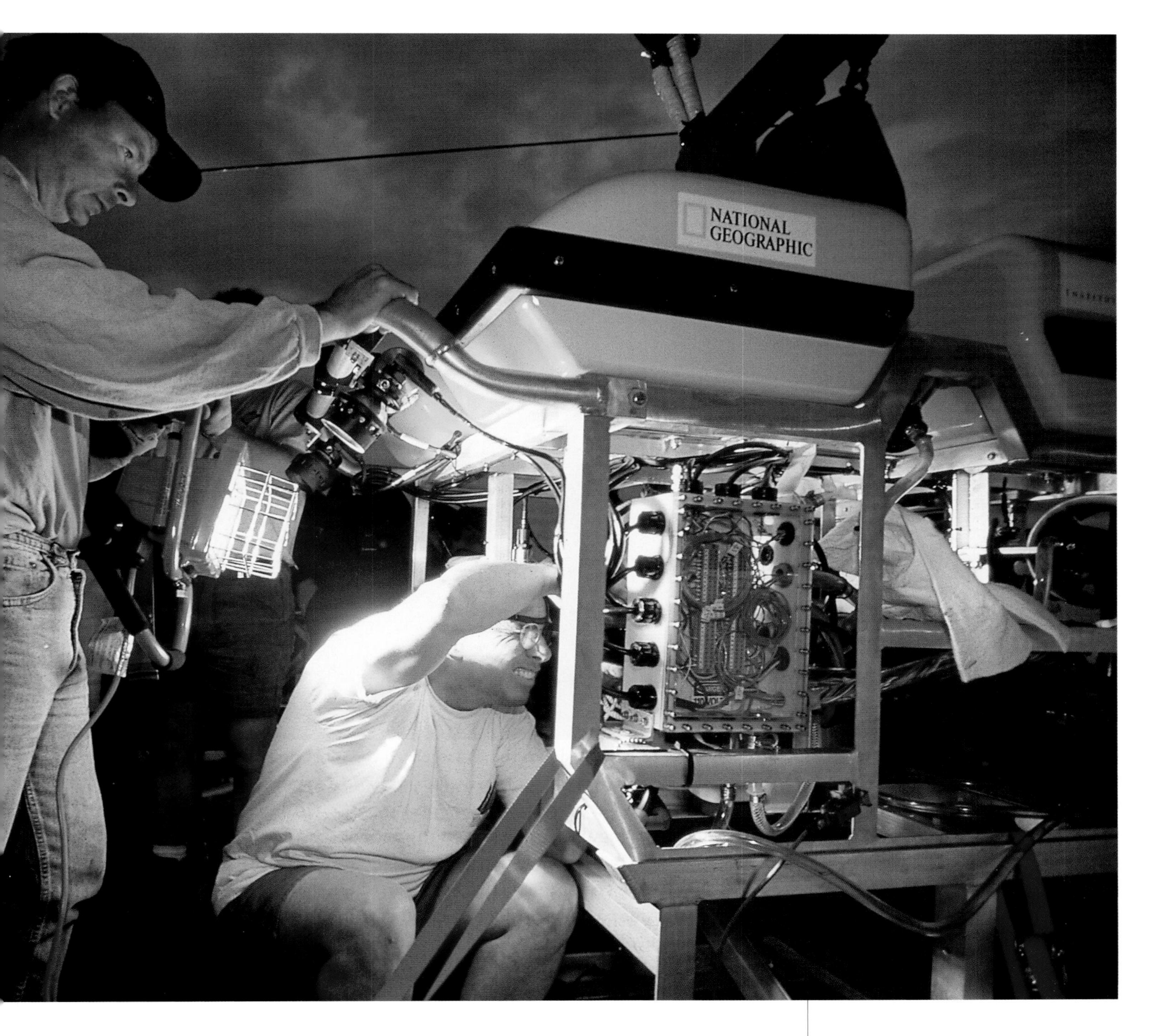

possibility, I explained. "But that's why you have a plan," I said with a grin, "so you can throw it away. You've got to be reactive."

The new watch took over their stations in the control room eagerly. As Tito Collasius worked the *Argus* winch, lowering the vehicle into the relatively shallow water, the control room became crowded with the usual curious off-watch spectators. I didn't have the heart to order them to their bunks for needed sleep on this first *Argus* dive. Because the vehicle is a towed sled, not an ROV, we had to maintain a fair altitude above the site or it would skid into the bottom. So the video image on our monitor was somewhat obscure. But I certainly planned to inspect this target in great detail as soon as *Little Herc* pilots Martin Bowen and Craig Elder arrived on

Dr. Fred Hiebert (on ladder), Jennifer Shadel Smith, and archaeologist Owen Doonan, all of the University of Pennsylvania, excavate the strata of the Sinop fortress. Artifacts revealed that the port had been active for millennia.

and swept back over the site, surveying the amount of wood and organic material available. "We need a piece of wood for carbon 14 so we can date it," I told the assembled crowd. After all, this was science, not a spectator sport.

My mind was buzzing with speculation and contradictory ideas. If we were in anoxic water, why were there little gray mullet making a meager living on the bottom around that wood? Obviously the water was oxygenated now. If this were the case over 7,500 years, wood eaters large and small would have consumed those logs and reeds long ago.

I forced myself to think rationally. We knew that the deeper layer below 170 meters (552 feet) was completely anoxic. After we towed the *DSL-120* in this depth, its iron clump weight came back to the surface dark black, stinking of rotten eggs (hydrogen sulfide)—the crew christened it "Tetanus." This unpleasant chemistry confirmed a virtual absence of free oxygen from 170 meters all the way to the Black Sea's bottom, 2,000 meters (6,500 feet). And we'd learned from local fishermen that the upper 85 meters (276 feet) were well oxygenated, providing rich catches of mackerel and flounder. Those same fishermen had spoken of a previously little known mixing layer that began at about 85 meters and extended down to 170 meters. This layer was dynamic, at times devoid of life when water from the anoxic layer welled up; but, when well-oxygenated water prevailed, free-swimming fish like these mullet could eke out a living. What was important about the discovery of this middle layer was that, unlike free-swimming fish that can escape the influx of the poisonous anoxic water, fixed wood-boring mollusks couldn't. If they had ever taken root in this layer, they would not have survived very long. When we had first considered the search for potential evidence of Stone Age inhabitants on the shallow coastal shelves of the Black Sea, we'd never given wood a thought because we were certain worms would have consumed it. Now, gazing at what appeared to be beams hewn by stone tools, we were in awe that these artifacts might well have survived.

We would need a wood sample for radiocarbon dating, but some of these supposed artifacts were no doubt driftwood a lot younger than 7,500 years old. If we got permission from the Turkish government to take a sample, we'd have to try to retrieve a piece of wood from under one of those stone blocks, trying to snag the oldest wood at the site. Unfortunately, we didn't have *Jason's* strong manipulator arm to lift one of the hewn logs.

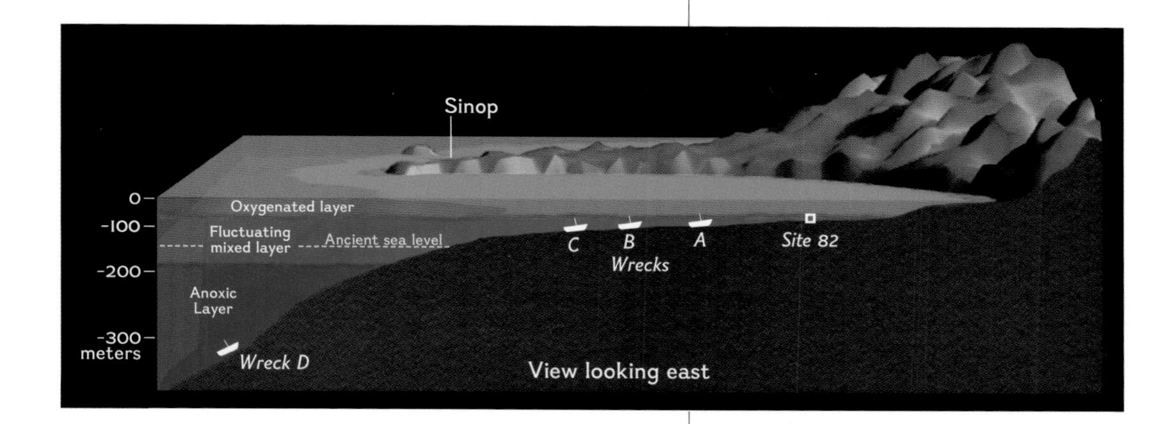

This computer graphic shows the depths and relative positions of the important sites discovered during our Black Sea 2000 expedition. Site 82, the possible Stone Age structure, and the cluster of shipwrecks are in shallow water. But the intact late-Roman ship (Wreck D) lies in the oxygen-depleted depths.

As we left the site, I turned to Fred Hiebert. "All right, Fred, tell me what we were looking at."

"I think we were looking at the four sides of a house." He gestured with his squared hands. "That might have been made of the traditional construction technique along the Black Sea coast, of mud packed around sticks with beams holding it all up."

"So what age period are we in?"

Fred shook his head. "That, we're going to have to do with radiocarbon dating. They made houses like this from 7,000 B.C. all the way up to 3,500 B.C."

"So carbon 14 on a piece of wood...."

"...is going to be big time," Fred said, completing my thought.

We'd have to choose the right piece or pieces of wood from this site, and be lucky enough to avoid contamination with modern waterlogged driftwood that had washed down from the coast and been swept along by undersea currents.

I sent a report to my sponsors at National Geographic Society headquarters in Washington, D.C., being very specific and careful in my wording, yet also trying to describe the site in some detail—it "consists of a single structure...most likely a house or shelter" that lay in a rolling, submerged countryside with a large river and tributaries meandering over its surface. I noted that there were logs and other artifacts of wood or stone clearly shaped by human hands found among the apparently collapsed stone matrix of the structure.

We were still at an early stage of our exploration of the site. But now we needed Turkish government permission to recover wood samples, and the best way to hasten that process was to go public with what we had found to date. This would ignite the media. The press would want to know if we'd discovered a Stone Age site. That would be their story; reporters wouldn't be happy with the "most likely" qualifier in my report.

Two nights later we used *Argus* and *Little Hercules* to zero in on two grape-cluster sonar targets, which were on a sloping bottom, falling away from about a hundred meters (325 feet). Tito Collasius was "flying" *Argus* with the winch control, and Martin Bowen controlled *Little Hercules* with his joystick console. The linked vehicles navigated on *Argus's* search sonar, which probed ahead. On the blue, pie-shaped wedge of the sonar monitor, the pink finger of the search beam swept relentlessly, revealing a jumbled target. A moment later, the video screen filled with the sharp image of heaped amphorae.

Radiocarbon dating revealed the smaller wooden specimens we retrieved from the site were recent. There is evidence Site 82 may be a Stone-Age structure submerged in the Great Flood, including this hand-hewn log and nearby stone block.

"Wait a minute," I said. "There's the wreck."

"Look at the size of that," Katie Croft gasped.

Already Katie was using an electronic calibrator to gauge the length of the exposed amphorae heap. Seventy-six feet was her preliminary estimate.

But I was more intrigued by the exposed deck and hull planks that lay reasonably intact among the stacks of long, carrot-shaped amphorae. Even though this wreck was in relatively shallow water, only a hundred meters, there still seemed to be some exposed wood that had evaded the hungry seafloor organisms. Obviously the phenomenon of the anoxic mixing layer was at work here.

Later that night we encountered a similar wreck. Nautical archaeologist Cheryl Ward studied the amphorae images intently, then nodded her head decisively. "Late Roman Empire," she said. "Fourth century A.D."

"All right," I conceded. "Not bad. Just a little too recent."

Late on September 12, I finally decided to get a full night's sleep while the crew used *Argus–Little Hercules* to inspect promising possible shipwreck targets in deeper anoxic water. I awoke to find all those targets had proved to be natural rock outcrops. But Cheryl Ward was fascinated with the video *Little Herc* had taken of one of these outcrops the night before, which closely resembled the material Fred was tentatively calling wattle-and-daub construction on Site 82. She thought it might be possible that this new site was actually a Paleolithic structure made of cut stone blocks. I compared the images of the two sites. Maybe, I thought. But the new site might not necessarily be a structure. If you were going to build with stone blocks, you would have to quarry that stone somewhere nearby. What I saw in these latest images reminded me of the sandstone I'd seen in the road cuts driving between the airport at Samsun and Sinop.

"Okay, Cheryl," I said, "I think we should review our entire database starting from the very beginning." I felt this process was critical; with the story breaking in the media, all the expedition scientists had to agree on certain fundamental facts and not go beyond them. We knew the press would want to link our discovery to the story of the Great Flood, but we could not be certain there was such a link, given what we had seen so far. And wishful thinking was not science. Cheryl, Fred, and I sat down in my cabin and spent a long day reviewing what we had and had not found at Site 82. We had not discovered a shipwreck. Ships were not made with rectangular stone block foundations, and the location of the site on a hilltop overlooking a river valley matched Paleolithic habitations that Fred had excavated on the nearby shore. After a long day of discussion, we reached fundamental agreement: We had

found a site 330 feet (101 meters) beneath the surface of the Black Sea that suggested human habitation before the area was covered by water. The newspapers and TV shows would carry this much further, of course, but this was as far as we could go on a firm scientific basis.

For the next several days we ran sonar searches for targets in increasingly deep water while we awaited permission to collect wood samples at Site 82. *Argus* and *Little Hercules* performed beautifully down below 350 meters (1,137 feet). We were ready to hunt for ancient ships in deep water, to paraphrase Willard Bascom. So we used the *DSL-120* to sweep north at depths from 250 to more than 500 meters (812 to more than 1624 feet).

It was soon time for our final personnel transfer, changing out the academic and technical staff that had to return to their institutions. On the way back to Sinop, we got word that the Turkish government had authorized our request to recover wooden samples from the archaeological sites. But we had to wait until the local museum director could be present aboard the ship on the 20th.

The director came onboard in the late morning of the 20th. By early afternoon, the wind was howling from the east and the seas were building. I frowned, wondering how Martin Bowen and Craig Elder were going to handle *Little Herc's* manipulator arm with the newly fashioned "Deep Scoop" recovery tool if that swell was snatching at the ROV's cable. But by the time we got to the site, the sea had calmed, and I made the decision to recover the wooden objects. Everyone in the

In *Northern Horizon*'s control room, our expedition crew looks on in wonder as *Little Hercules* pilot Craig Elder (right) tweaks the ROV's joystick, bringing it closer to Site 82.

***Following pages:* The distinctive, carrot-shaped amphorae indicate a late-Roman Empire shipwreck, here one discovered off the Turkish coast. Most of the hulls of these ships had been consumed by woodborers, but the shapes of the vessels were clearly evident.**

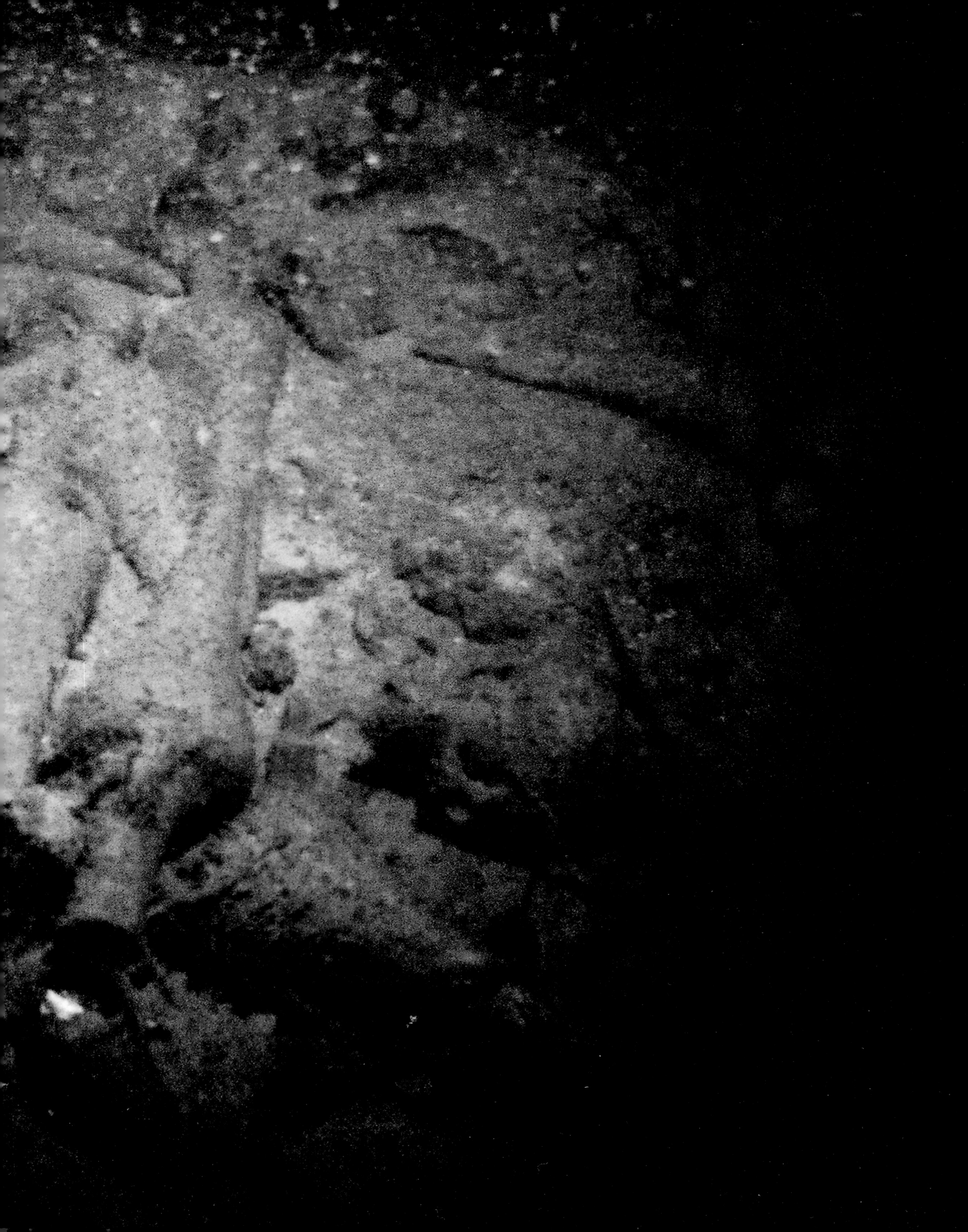

The lights of *Little Hercules* (photographed by *Argus*) penetrate the stygian darkness around the ancient intact shipwreck on the floor of the Black Sea. The discovery proved that the oxygen-depleted waters must be a virtual museum of the deep, since ships have sailed the stormy Black Sea for millennia, and countless vessels have sunk.

control room was tense as Martin gingerly nudged past a block of sandstone, deftly keeping the ROV in place with its aft thrusters so as not to stir up sediment at the forward end. He was after a wooden wedge that clearly bore evidence of human working. As we held our breath, Martin recovered the wedge with a quiet, "Got it."

Because we didn't have our old elevator system with us, we had to raise and lower *Little Hercules* each time we collected a wood sample from the site. All night, the painstaking process proceeded, with Martin working his four-hour watch, then Craig Elder taking over. We stored our small samples in seawater inside Tupperware boxes that Cathy Offinger had bought in Sinop—shades of our impromptu Galápagos Rift hydrothermal vent biology sampling in 1977.

On September 21, we dropped the museum director back in Sinop and headed north, back into deep water to inspect our promising possible shipwreck targets. As we ran our long survey lines, the wind rose steadily, increasing to a gale from the east northeast. By 6:00 p.m. on September 23, the wind was blowing a steady 40 knots and the sea was a wilderness of confused swells and whitecaps. In these conditions, it would be too dangerous to try to recover the *DSL-120* and launch the *Argus–Little Hercules* package to visually inspect any possibilities.

But the weather-enforced hiatus gave me a chance to carefully analyze the previous days' sonar-run database. Unfortunately, I had to return to the States for a JASON Foundation board meeting, so Dwight would have to take over the last few days of the expedition to exploit the anoxic depths. After we docked in Sinop, I met with Dwight and we carefully went over my analysis of the sonar targets of the last *DSL-120* run. I had narrowed the target list down to six, the most promising of which was labeled "Anoxia TGT #52." It lay in 350 meters (1,137 feet) of water, north of the coastal shelf. The tiny gray smudge could have been an ancient ship that had sunk on the treacherous passage between Sinop and the Crimea centuries or millennia before.

"Okay, Bob." Dwight conformed. "That's the one we'll hit first."

At midnight I watched rather wistfully while the *Northern Horizon* sailed out of the harbor in search of the elusive well-preserved ancient ship in the anoxic deep water of the Black Sea.

After a two-plane hop to Paris, I called home from Charles de Gaulle Airport and my wife Barbara proudly announced, "They found a ship. Its hull is completely intact. It even has its mast." I was stunned, but didn't learn the details until the next day when Dwight sent me a long e-mail with attached images of this amazing new shipwreck.

The single-mast rigging and hull design, unchanged for centuries, here on a bas relief of a Classical merchant vessel on the sarcophagus of a Roman mariner on display in the Sinop museum, were probably similar to the ships carrying amphorae we discovered near the coast.

After leaving Sinop, they'd reached the site late the next morning and then launched *Argus–Little Hercules*. As Dave Wright manned *Argus'* winch and PJ Bernard worked the *Little Herc* console, the linked vehicles crept forward toward a vague target on the scanning sonar's blue screen. The first thing the control room watch saw on the video monitor through the drifting marine snow in the floodlights was a tall pole rising from the bottom gloom. "Is that a tree?" someone muttered. PJ banked the little ROV for a closer look. The video camera revealed what appeared to be an upright tapered tree trunk, then panned down to display the ghostly image of wooden posts and timbers. A crew member ran to wake Texas A&M nautical archeology graduate student Kathryn Willis, who had gone off watch. With both Cheryl Ward and Fred Hiebert having left the expedition for home institutions, Kathryn, who was on her first major international cruise, now had unusual responsibilities thrust upon her.

When she reached the control room, the watch was deeply puzzled over what they'd found. "Are those timbers?" Kathryn asked, trying to keep her voice even. They had encountered jumbled driftwood closer inshore and no one dared voice the mounting hope they all felt: that this weird assembly of wood more than 300 meters (1000 feet) below the surface of the Black Sea was in fact the ancient ship we sought. But after a slow, pain-staking reconnaissance, *Little Herc's* video camera returned to the upright tree.

"I think it's a mast," Kathryn said.

"I agree," Dwight concurred.

The watch reached consensus that they were looking at the mast of a sailing ship. The mast stood 11 meters (35 feet) tall, leaning slightly forward and tapered at the top. As the camera zoomed in, the crew gasped as they saw remnants of rope still cross-knotted near the top of the mast.

"We felt deep emotion," Dwight later told me, describing the reaction of those around him in the control room. "Was this was the prize we had all been dreaming of?"

In the glare of *Argus'* floodlights, the entire 12-meter (39-foot) hull of the ship, buried up to the deck in the dead anoxic mud, came into view. They could definitely see the intact four-meter (13-foot) width of deck planks visible beneath the veneer of gray-brown sediment. There wasn't much visible of the bow section because it was more deeply covered, but there was a pair of vertical beige stanchions with neatly carved mortise notches standing near the mast. Farther aft stood a second pair of similar stanchions, topped by square tongues. These four posts had apparently formed a rectangular framework that had either supported heavy spars when the sail was lowered or formed a deck shelter, or had possibly served both functions. The notched sternpost was obvious, as was the stout rudder support rising from the starboard bulwark, where it had once gripped the now-missing steering oar, possible mute evidence to the violence of the storm that had sunk the ship. A thick wooden spar with a rounded tenon end lay on the deck among broken timbers and sprung hull frames that had been smashed by this heavy timber as it came crashing down. *Little Herc's* video also revealed lighter curved wooden spars lying on the deck that were probably part of a triangular lateen rig. This debris could have accounted for the absence of lines and sailcloth: The vessel's rigging had probably been shredded by a squall before the heavy cross-spar fell. But several planks were scored by worn grooves of ropes and the grain of the wood was visible on others.

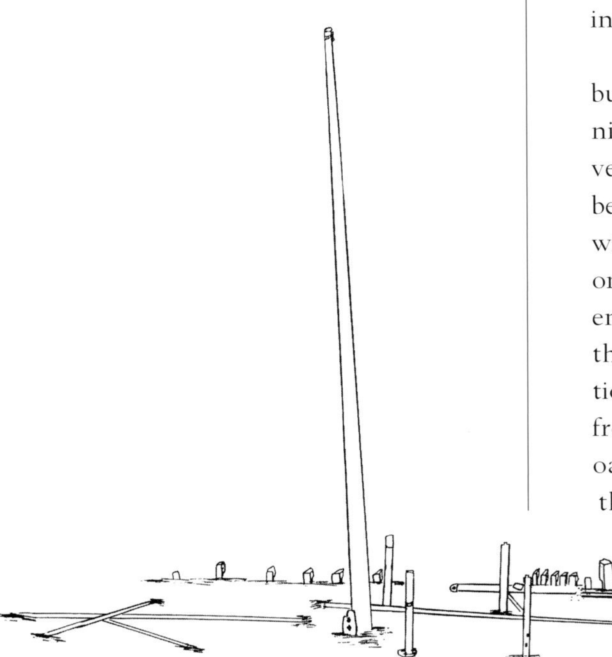

The mast of Wreck D stands 35 feet, and the stanchions form a square framework that might have housed a deck shelter. The notched sternpost and the stout steering-oar support on the starboard bulwark are obvious.

"We didn't see a single speck of metal on the wreck," Dwight said.

Still, to Kathryn Willis, this initial video inspection did not prove that the wreck was ancient. She knew the shipbuilding traditions of the Black Sea had changed little over the centuries. This vessel could be a fishing boat that had foundered in the early 1900s or a Greek ship from the time of Homer. But when Cheryl Ward later reviewed the close-up videotape of the ship, the only fastenings she found were primitive wooden "treenails," pegs driven through drilled holes. "This definitely was not a modern vessel," she told me. "Yet no archeologist has ever seen an ancient shipwreck in such a perfect state of preservation. It looks as if it has just left the dock."

Continuing their inspection, the expedition crew found more fascinating details. Amidships, there were a pair of stout posts on the port and starboard sides mounting a thick cross beam that could have been the main frame of the hull, which had also possibly formed an edge of the cargo hold. A circular object appeared in the lights, perhaps the mouth of an amphora. Using the ROV's thrusters, PJ tried to dust the sediment off for a better look. But the image only clouded with thick dark mud that took a long time to settle.

Dwight requested the Turkish government's permission to retrieve a wood sample for carbon-14-dating. But while the crew waited, a rainy gale blew up, producing marginal conditions for continued work with the submerged vehicles. Dwight was determined to press on and conferred with Kathryn, Martin Bowen, and Tito Collasius to devise a strategy to retrieve a wood sample in the least destructive manner. Martin and Tito rummaged around in the tool van and found a corker, a sharply tapered steel tube normally used to take flesh samples from large fish. Tito welded the corker to a steel rod and mounted it high on Little Hercules. The next day the Turkish authorities granted permission to take the wood sample. Kathryn was worried about accepting the responsibility of selecting the best spot on the wreck to take the sample: This would involve ramming the ROV head-on into the ship so that the corker's tube could fill with wood. But would the crude tactic damage the precious shipwreck?

She chose the sturdy rudderpost on the aft starboard side. Battling the swell and blinding rain, the crew launched the vehicles. Twenty minutes later, Martin Bowen tested the rudderpost's strength by nudging it with the ROV. The wooden stanchion did not budge. Then he reversed *Little Herc* and sent it full ahead to imbed the corker in the wood. Three times he repeated the maneuver. On the third attempt, the corker stuck so deep that the ROV was held immobile, no matter how hard Martin reversed the thrusters. Finally, amid a cloud of blinding sediment, *Little Herc* broke free. Kathryn was almost sick with anxiety waiting the 20 minutes for the vehicles to be recovered. Had they retrieved their precious wood for carbon-14-dating?

"We've got a good sample," Tito told her, bending to examine the corker.

This sampling was the only actual physical contact our expedition made with the ship. When *Argus* and *Little Hercules* finally lifted their floodlights clear of the anoxic bottom, the debris-strewn deck and gracefully carved mast seemed to sink once more into the eternal night of the abyss, just as they had after that terrible storm so many centuries ago.

An artistic rendition of Wreck D, based on Cheryl Ward's diagram, portrays the curved wooden boom attaching the triangular sail to the mast found lying on deck, possible evidence the ship had lost its rig in a storm.

Fred Hiebert coordinated the radiocarbon dating of all the expedition samples, using the ultra-sensitive accelerator mass spectrometer (AMS) technique. The results were somewhat mixed. As I had feared, the small wood samples recovered from Site 82 tested out as modern, less than 200 years old. They were contamination that had drifted along the bottom with the plastic, trash, and soft-drink cans we had also encountered. Unfortunately, we had not been able to retrieve a sample from one of the large logs that showed the best visual evidence of having been hewn by human hands. That would be a goal of a future expedition. But Fred was able to have a chemical analysis made of a mud sample we'd taken from among the stone blocks of the site. The material proved to be of a different composition than the seafloor sediment outside the site. This was evidence that the mud had been part of a wattle-and-daub wall, a further indication that we had in fact discovered a Paleolithic structure. And the mollusk shells we'd collected from the site were all saltwater species, indicating the location had never been submerged in the ancient freshwater lake. This meant the rectangular structure had stood on a hilltop above a river valley when the area was dry land.

The radiocarbon dating on the deep, perfectly preserved wreck confirmed what we had learned from Cheryl Ward's analysis of the spectacular video images. It was very old: The ship dated between A.D. 410 and 520—the late-Roman/early-Byzantine period. To discover a ship approximately 1,500 years old in virtually

flawless condition—including the knotted rigging lines on the masthead—was success far beyond anything I had hoped for on this expedition.

We had confirmed Bascom's theory that the Black Sea held the promise of preserving more history in its depths than any other location on the planet. Given the central maritime position of the Black Sea from the Bronze Age through the Roman and Byzantine Empires, I knew there had to be hundreds or even thousands of similar wrecks lying silently in its dark anoxic depths.

Putting these discoveries in perspective, I realized that my earlier exploration

Our Black Sea 2000 expedition succeeded beyond all expectations. We will return to explore again, hunting for more intact shipwrecks. I am confident we will find even older vessels, perhaps dating to the Bronze Age and mythical heroes such as Jason.

of the Mid-Atlantic Ridge and the discovery of the Pacific hydrothermal vents with their unique ecosystems had been more scientifically significant than this Black Sea expedition. But viewing the submerged foundation stones of a possible Paleolithic structure and the intact wooden hull of an ancient ship in the Black Sea's lifeless abyss through the telepresence of *Little Hercules'* video eye had been experiences that had reached deep into my psyche.

I felt the satisfaction of successful exploration. We had come seeking knowledge, and we had found it. And I knew we would return.

Epilogue

Jason, Jr., **illuminated in the back glow of** ***Alvin*****'s lights, explores the stalactites of rust cascading down the hull plates of** ***Titanic*****'s bow. During my second visit to the famous wreck in July 1986, I left a memorial plaque to the victims of the disaster. That last dive to** ***Titanic*****'s grave is one of the most haunting images I have retained during four decades of exploration.**

As an Explorer-in-Residence at the National Geographic Society, people often ask me about the most memorable images I have retained from scores of ocean expeditions over four decades.

Certainly deep ocean archaeology conducted with *Jason* has been an inspiring experience. I will never forget holding in my palm the delicate terra-cotta oil lamp recovered from the grave of the ancient Roman ship we named *Isis* or the Phoenician incense chalice we recovered from *Elissa*.

And the sight of the ghostly parade ground of drowned German sailors' boots near the sunken *Bismarck* remains a haunting image.

But perhaps the picture from the ocean floor that has lingered with me the longest is a view of *Titanic* I had on July 21, 1986, during our second season on the wreck site. I visited the ship in *Alvin*. Our pilot, Ralph Hollis, took the submersible on a close inspection of the stern, during which I used the manipulator arm to deposit a memorial plaque to the victims of the sinking. On this deck so many of them had clustered before the ship's last vertical plunge into the icy, starlit Atlantic.

When Ralph dropped descent weights, *Alvin* gained positive buoyancy, and we began to rise toward the surface more than two miles above. I watched the sunken liner slip from the glow of our floodlights as the sea's eternal darkness returned to shroud the wreck.

Over the years since then, I have dived in submersibles to visit the lost liners *Lusitania* and *Britannia* and to sunken naval vessels such as the huge Japanese battleship *Kirishima* resting on the haunted bottom of the Slot near Guadalcanal. Each experience reinforced my conviction that the seafloor is a vast, unspoiled museum.

I was not aboard the *Northern Horizon* when last year's Black Sea expedition made the stunning discovery of the perfectly preserved ancient ship in the anoxic depths. But it is fitting that a new generation among my expedition team was clustered in the control room when the first ghostly images of that tilted mast appeared on the video monitor through the marine snow.

Exploration is not an individual enterprise but a continuing human quest for knowledge that links my colleagues and me in an unbroken chain of discovery reaching back to the Bronze Age mariners, the Phoenicians, and the far-ranging Polynesian voyagers.

Ocean exploration continues. And now the precious knowledge with which explorers return to port is not the Golden Fleece or word of a fertile island beyond the horizon, but a fuller understanding of our intriguing past and the nature of our unique planet.

Peterson, Mendel. *The Funnel of Gold*. Boston: Little, Brown and Company, 1975.

Pond, Seymour Gates. *Ferdinand Magellan, Master Mariner*. New York: Random House, 1957.

Sontag, Sherry, and Christopher Drew, with Annette Lawrence Drew. *Blind Man's Bluff, The Untold Story of American Submarine Espionage*. New York: BBS Public Affairs, 1998.

Spate, O.H.K. *The Spanish Lake*. Minneapolis: University of Minnesota Press, 1979.

Waldman, Carl, and Alan Wexle. *Who Was Who in World Exploration*. New York: Facts on File, Inc., 1992.

Weckler, Jr., J.E. Polynesians, *Explorers of the Pacific*. Smithsonian Institution War Background Studies, Number Six. Washington: The Smithsonian Institution, 1943.

Welker, Robert Henry. *Natural Man, The Life of William Beebe*. Bloomington: Indiana University Press, 1975.

Williams, Frances Leigh. *Matthew Fontaine Maury, Scientist of the Sea*. New Brunswick, N.J.: Rutgers University Press, 1963.

Zweig, Stefan. *Conqueror of the Seas, The Story of Magellan*. New York: Viking Press, 1938.

ARTICLES AND JOURNALS

Ballard, R.D., A.M. McCann, D. Yoerger, L. Whitcomb, D. Mindell, J. Oleson, H. Singh, B. Foley, J. Adams, D. Piechota, and C. Giangrande, "The Discovery of Ancient History in the Deep Sea Using Advanced Deep Submergence Technology," *Deep-Sea Research*, September 2000.

Ballard, R.D., and J.F. Grassle. "Return to Oases of the Deep." *National Geographic* 156(5):680-705.

Ballard, R.D., T. Holcomb, and T.H. van Andel. "The Galapagos Rift at 86(W:3. Sheet Flows, Collapse Pits, and Lava Lakes of the Rift Valley," *Journal Geophysics. Res.* 84(B10:5407-5422

Ballard, R.D., J. Francheteau, T. Juteau, C. Rangan, and W. Normark. "East Pacific Rise at 21(N: The Volcanic, Tectonic, and Hydrotheermal Processes of the Central Axis." *Earth Planet. Science Letter*. 55(1):1-10.

Ballard, R.D. "The Exploits of Alvin and ANGUS: Exploring the East Pacific Rise." *Oceanus* 27(3):7-14.

Ballard, R.D. "High-Tech Search for Roman Shipwrecks." *National Geographic*, April 1998: pp. 32-41.

Ballard, R.D. "NR-1. The Navy's Inner-Space Shuttle." *National Geographic*, April 1985: pp. 450-458.

Bellwood, Peter. "Ancient Seafarers." *Archaeology*, March/April 1997, Vol 50 No 2.

Booth, William. "Early Migrants May Have Come by Land and Sea." *Washington Post*. September 6, 1999, page A13.

Corliss, J.B., J.A. Baross, and S.E. Hoffman. "An Hypothesis Concerning the Relationship Between Submarine Hot Springs and the Origin of Life on Earth." Oceanol, Acta, 1981. *Proceedings 26th International Geological Congress, Geology of Oceans Symposium*, Paris, July 7-17, 1980, 56-69.

Fladmark, Knut R. "The Feasibility of the Northwest Coast as a Migration Route for Early Man." *Early Man in America, from a Circum-Pacific Perspective*, Alan L. Bryan, editor), Occasional Papers No. 1, Dept. of Anthropology, University of Alberta, Edmonton, 1978. P. 119.

Fladmark, Knut R. "Times and Places: Environmental Correlates of Mid-to-Late Wisconsinan Human Population Expansion in North America." *Early Man in the New World*. (Richard Shutter, Jr. Editor), Sage Publications, Beverly Hills/London: 1983.

Kawaharada, Dennis. *The Settlement of Polynesia, Part 1 & 2.* Http://leahli.kcc.hawaii.edu/org/pvs/migrationspart1.html, University of Hawaii, Department of Anthropology.

Kourakou, Stavroula. "Charting the Course of Pine: Resin Wine in Golden Mycenae." *Kathimerini*, Athens. Nov. 13-14, 1999.

Meltzer, David J. "North America's Vast Legacy." *Archaeology*. Jan/Feb. 1999, p. 51.

Rose, Mark. "Beyond Clovis, How and When the First Americans Arrived." *Archaeology*. Vol 52, Number 6, Nov/Dec 1999. Online: http://www.archaeology.org/9911/etc/books.html.

Rose, Mark. "The Importance of Monte Verde." *Archaeology*. Online Features: "Monte Verde Under Fire." Http://www.archaeology.org/online/features/clovis/rose1.html.

Schuster, Angela M.H. "Ancient Mariners Found in Peru." *Archaeology* Online News, http://www.archaeology.org/online/news/peru.html.

"Phoenician Ships and Signs of the Biblical Flood." *Undercurrents*. Vol. 8 No. 4, Fall 1999.

Wilford, John Noble. "Chilean Field Yields New Clues to Peopling of Americas." *New York Times*. August 25, 1998.

ADDITIONAL ONLINE RESOURCE:

Internet information on the JASON Project is available at:

http://www.jasonproject.org.

INDEX

Boldface indicates illustrations.